Table of Contents	2
List of Figures	4
Preface	6
Disclaimer	7
- Jargon	7
- Terminology	8
a) Minerals	8
b) Recombination	9
c) Cybernetics	9
- Citation	10
- Brands and Trademarks	10
- Copyrights	10
[I] 'Kingdom Minerals', Why Abandoned?	**11**
Story time!	11
[II] Homocybernetica: The 2% 'Chimpanzee-nct'	21
2.1 A unique extended-phenotype	21
2.2 The machine maker & companion	25
2.2.1 Education & mass-communication	27
2.2.2 Cybernetics	29
2.2.3 Culture	40
2.2.3.1 The 'ET' Culture	40
2.2.3.2 The Cybernetics recombination: Machines as living-systems	45
2.2.3.3 Culture as a genomic supplement	51
2.2.3.4 Expressing life dynamics: A holistic simplified model	52
2.3 'Wisdom': The Cybernetic Selection	57
2.4 The Homocybernetica: A new Kingdom of life; We	

	are not just another species	65
[III]	**Cybernetica: Machines Beyond Association**	**69**
	3.1 The life story of a machine	73
	3.2 The Cybernetic tripartite: A close up!	94
	3.3 Machines on the route of autonomy	101
[IV]	**The future of humanity & machines**	**103**
	4.1 Co-evolution; the past, present, and the future	103
	4.2 The 'black swan'[41] and the future of robots	111
	4.3 Possible scenarios	115
	Scenario 1	115
	The survival of the Homocybernetica:	115
	Scenario 2	116
	The prevalence of Cybernetica	116
	Scenario 3	119
	the Praying-Mantis Scenario	119
	The Darwinian 'Virtue' Scenario	122
	Scenario 5	124
	The extinction of machines by viruses or Cyber wars:	124
	Scenario 6	125
	Total chaos and extinction of life	125
	Scenario 7	125
	The Red Swan Scenario	125
	4.4 CONCLUSION	**127**
	Reference List	132

THE CYBERNETIC RECOMBINATION

The Biology Of

Technology

Revisiting Linnaeus' Kingdom Minerals
Gihan SOLIMAN

Author: Gihan S. Soliman, BA, PGCE, MSc, RSci

To cite this work:

Soliman, G. (2019) The Cybernetic Recombination, The Biology of Technology; Revisiting Linnaeus' Kingdom Minerals. 4th edition (Colour).

Copyrights @October 2019 - registration No. 284731172

1st edition October 2019

First published in October 2019 under the title of : The Homocybernetica: a New Kingdom of Life; Revisiting Linnaeus' *Regnum Lapideum.*

List of Figures

Figures	Title	Page
1.1	The extended phenotype of *Homo sapiens* versus that of an animal (e.g. Beaver)	24
2.1	hIEM: A Cybernetic model of the genotype-phenotype-extended phenotype correlations and dynamics of a lower animal	56
2.2	A Cybernetic model of the genotype-phenotype-extended phenotype correlations and dynamics of the Homocybernetica	57
2.3	Parallelism: Homocybernetica versus a living cell	63
2.4	A Cybernetic expression of Kingdom Homocybernetica	68
3.1	My car or Xmas?	78
3.2	Example of machine evolution - Citroen.	82
3.3	The 'Yeast Parable'	86
3.4	A proposed taxonomy of a machine as an associated organism. e.g. Peugeot 108	86
3.5	Two supra domains of life	
3.6	The Homocybernetica: A physical-biological organism	92
3.7	The *Systema Naturae* revisited	94
3.8	The emergence of 'cuteness' in robots	99
3.9	The evolution of social skills in robots - Sophie & Hans	101
4	A timeline of evolution of life on earth..	109

Preface

Are modern biologists behind the curve in identifying life?
This work provides evidence that such is the case, with a call for liberating biology from the shackles of protein, by reinstating Linnaeus' abandoned 'Kingdom Minerals' as a supra-domain of life. By revisiting Linnaeus' & Darwin's seminal works, in light of new scientific discoveries, we conclude that the faculty of Cybernetics is a discounted force of nature, which is the force of re-aligning matter purposefully, or the Third Recombination. Two recombinations construct bottom-up living formations, while Cybernetics constructs top-bottom living systems. This work, therefore, provides evidence from natural history and academic research that technology *is* a form of life and that *we* are not just another species of mammals. The proposition that machines live and behave as species, in association with humans, is not presented as a piece of science fiction, prediction or clever projection; but as the *status quo* - as is and has been for thousands of years. The book proposes a correction of the current taxonomic orders to include two new mega life-groups namely the 'Homocybernetica' - comprising the tripartite; humans, machines and culture, and the 'Cybernetica', comprising the novel AI technology *and,* surprisingly, the neglected ancient natural mineral living-cycles. The proposed taxonomic order Homocybernetica closes the loop for systematics by being the missing link between the biological and physical realms known collectively as Nature, and a nod of convergence between animals and the rising intelligent robots. This book shows that technology as it evolves is neither controllable nor predictable but is perhaps only possible to slow down long enough for potential co-evolution and survival. Seven possible scenarios for the future of humanity with machines are, hereby, presented.

Disclaimer

Jargon

One of the biggest challenges of interdisciplinary research is addressing jargon. As a result of dissecting human knowledge into tightly-boxed fields, one term may refer to completely different, or even conflicting, thing(s) as it crosses domains. An example of this is the use of the term 'organic' - critically relevant to the research in hand, in different domains of knowledge (summarised from Oxford dictionary [56] and others) :
Organic (Chemistry): Containing carbon and from biological origins.
Organic (Food): Grown with no synthetic fertilisers.
Organic (General): Pertaining to life.
Organic (Systematics): Consisting of different parts that are all connected to each other.
Organic (Organisations): The law by which an organisation functions.
Organic (Change) Happening naturally over time without being forced or planned.

The author, therefore, uses such expressions cautiously followed by an indicator of the field (e.g. organically (Chem, Biol, Sys, Life, etc.)). Additionally the following expressions are either avoided or used sparingly:
Biotic, abiotic, natural, naturally-occurring, spontaneously-occurring, and artificial, to eliminate any confusion that

might arise from the different uses of such expressions from one domain to another.

- **Terminology**

The author uses the following terminology – instead, to avoid speciality-biased interpretations - with apologies for their inevitable awkwardness:
Protein-based forms of life: Living-systems that once emerged in water, with a living cell as the fundamental unit of their structure. In *Linnaeus'* time, they were perceived (only) as animals and plants but by the advancement of knowledge and molecular biology, more taxonomic groups (kingdoms) have been identified such as fungi, bacteria, etc.
Mineral-based forms of life: Living systems in which minerals - on the pre-cellular, non-cellular, and beyond cellular level; engage in self-regulatory and self-replicating loops performing dynamic functions subject to evolution and natural selection, against an environment. Mineral-based forms of life in the context of this book are not necessarily protein free but rather freed from the *limitation* of protein in their composition. Such systems have been either overlooked or ditched by modern biologists despite having the markers of life, as demonstrated in the following chapters.

a) Minerals
The term minerals hereby is used in a cross-disciplinary

sense - as was also used by *Linnaeus*, in reference to the elements circulating systematically in nature in units (or recombinations) lower than, different from, beyond, or parallel to protein-based living cells such as rocks, mineable minerals, nutrients, and carbon atoms as they cycle in living organisms and fossils.

b) Recombination

Recombination (in biology) [59] refers to the process by which pieces of DNA are broken and recombined to produce new combinations of alleles. This recombination process creates genetic diversity.

Without actually calling it 'recombination', Linnaeus also (insightfully but prematurely) describes such a persistent process of elemental recombination driving geological formations such as rocks and minerals in nature, parallel to the sexual reproduction in plants.

Recombination (in cosmology) refers to formation of matter at 'the time of last scattering' or the 'apparition of light', 13.8 billion years ago when the temperature of the universe dropped enough to allow electrons to bind to nuclei and form the first neutral atoms.

Recombination (in Cybernetics) is introduced by the author hereby - in parallel to biological and cosmological recombinations, in reference to any *replicable* re-alignment of matter (minerals in the Linnaeus' sense) as to manipulate the natural flow of energy purposefully and *recursively* in observation of the laws of nature and the four fundamental forces/interactions.

c) Cybernetics

An approach exploring the systematics of 'steermanship' or 'self-regulation' in animals, machines, and organisations. The faculty of Cybernetics is the uniquely-human ability to observe, drive, and mandate self-regulation by editing system information.

- **Citation**

Due to the complexity of proposition, two types of references have been used: Peer-reviewed materials which contribute to the rationale of the paper, and reliable documentation, educational materials as well as media links to simplify the background knowledge to non-specialists without jeopardising the integrity of the discourse. An example is the structure of a living cell.

- **Brands and Trademarks**

Brands and trademarks depicted in this book are meant for illustration only. The author is not affiliated with, promoting, recommending, or advising against any of the businesses/products/brands mentioned in this book and has no financial interest with them whatsoever.

- **Copyrights**

No infringement of IP rights is hereby intended. If you

have concerns that your IP rights have been breached, please contact the author to have this rectified in a second edition with any due apologies.

[I]

'Kingdom Minerals', Why Abandoned?

Few decades ago, A Swedish bright scholar by the name Carl Linnaeus set the foundations of modern taxonomy in his work *Systema Naturae* (first published in 1735). The 10th edition of that same book[1] classified the 'natural system' into three 'kingdoms' (i.e. majour taxonomic groups). He named his dear groups, *'Regnum Animale'*[2] *(*Kingdom Animals), *'Regnum Vegetabile' (*Kingdom plants),* and *'Regnum Lapideum' (*Kingdom Minerals) [18][19][20]. This is known to have marked the birth of the binomial nomenclature still used for classifying and naming living organisms [30]. Taxonomy has, since then evolved and classification approaches have varied greatly - yet, Linnaeus is still regarded as the father of modern

---

[1] Published twenty three years later.
[2] In Latin

taxonomy [29], as well as the founder of modern ecology [11]. One of his kingdoms, however, couldn't survive long.

At this point, we need to remember that taxonomy is the heart of nomenclature which, in turn, is the backbone of life-science when it comes to conservation and academic research [36]. The application of the *Linnaeus*' binomial[3] nomenclature system [4] is subject to certain criteria that allowed the binomial system to evolve over time by expanding on naming the cellular protein-based forms of life, while the Kingdom Minerals w th all its rocks, minerals, fossils, and other elements circulating in nature has been banished.

To see the implication of such abandonment, let's take a

---

[3] A scientific name of an organism composec of two parts.
[4] Governed chiefly by the International Code of Zoo ogical Nomenclature (*ICZN*) [49] and the International Code of Nomenclature for algae, fungi, and plants (ICNafp) [48]

look at the Linnaean taxonomy used today to classify living organisms into 9 hierarchical groups:

Domains,
    kingdoms,
        phyla,
            class,
                order,
                    family,
                        genus,
                            and species

All pertaining to cellular living-organisations known as 'organic' (Life).

The contemporary human kind, for example, is classified as a species of mammals and is identified as the *Homo sapiens,* which is a combination of the genus and species classification in the following hierarchical order:

Domain: Eukaryotes;

   Kingdom: Animalia;

      Phylum: Cordata;

         Class: Mammalia;

           Order: Primate

              Family: Hominidae

                 Subfamily: Homininae;

                    Tribe: Homnini;

                       Genus: Homo

                          Species:

                          Homo sapien [25]

As for minerals, Linnaeus divided them into four majour groups: Rocks, Minerals, Fossils, and Nutrients, all emerging and developing through the natural recombination of elements parallel - as he believed, to the sexual reproduction in plants. Linnaeus never actually proposed that rocks or fossils or nutrients actually *copulated* or had any genitalia but that minerals underwent some sort of spontaneous re-assembly parallel to the biological recombination processes underpinning the diversity of plants. The exact mechanism of recombination, understandably, couldn't have been identified then given the rudimentary body of knowledge on elements of the time; Enough to mention that the periodic table[5] itself had not been established yet and many of the important elements such as Nitrogen had not even been discovered [40! Because Linnaeus' 'sexual'-based classification of minerals seemed way off the mark at the time, the concept was understandably rejected by both biologists *and* mineralogists alike and never been really revisited. One reason why it's never been revisited remarkably is the yawning gap in communication among science domains as their scholars

---

[5] completed by Mendeleev, D. (1869)

came to use different jargon and have their own biases. So basically, the correction of the biological error of abandoning a majour living system has already occurred but in different languages which biologists don't understand or speak, such as Cybernetics and some branches of Physics. As it turns out, the concept of (elemental) recombination is a cosmological given regardless of the mechanism and is as critical to studying the emergence of the universe [53] as it is to understanding biological complexity and diversity. Needless to say, that the elemental forces underpinning any biological recombination are mostly written in physics and chemistry jargons. This is an issue that Darwin clearly tapped on by pointing out that the mystery of life-emergence is beyond cells and entails a deep understanding of physics not yet available to humans at his time:

> .. science as yet throws no light on the origin of life. Who

can explain what is the essence of the Attraction of gravity? [7]

Unfortunately, by the time our understanding of cosmology somehow matured the 'Kingdom Minerals' had already lost its position as a system of life and been forgotten even by the *Linnaeians* especially after being already ripped apart among mineralogists, ecologists, paleontologists, naturalists, nutritionists, dieticianists, biochemists, chemists, physicists, and engineers; each to their jargon and narrowly-defined expertise; and being treated as 'inanimate'.

The importance of revisiting Kingdom Minerals (and subsequently the *Systema Naturæ)* arises from the desperate need to liberate biology from the shackles of proteins, as many mineral-based non-cellular and protein-free entities have been displaying the fundamental characteristics of life yet never found their way to the modern taxonomy[6]. The observation of the dynamic mineral cycles in nature - from a biological fresh perspective, may also help shedding some light on the cross-species interchange of genetic materials complementary to the hierarchical perspective known metaphorically as the 'tree of life' (see [31]) Such tree would otherwise have several missing links and is unjustifiably truncated at the roots while getting bushier and bushier especially with the acvancement of technology and molecular biology.

Ironically, the same advancement of technology that pushed Kingdom Minerals out of the system and expanded on molecular biology, is beginning to highlight

---
[6] (as applied in biology)

the fundamental flaw of such protein-based definition of life by discounting the behavior of 'Minerals'[7] as they circulate systematically in nature - such as in the carbon [1] [35], nutrients, and water cycles and as they undergo diverse sets of recombinations to form vital association with the human kind by the emergence of machines on the blue planet.

It might be difficult to accept that a social organisation or a machine made of 'inanimate' metal/minerals is actually a form of life but the reality is, nothing in this universe is inanimate from atoms to stars, all is in constant motion! If evidence demonstrates that minerals[8] recombinations behave like species in expression of given packs of

---

[7] Important to remember that Linnaeus 'Minerals are different from the minerals as defined by mineralogists today as they spanned rocks, mineable minerals, fossils, and nutrients.

[8] The term minerals is capitalized when referring to Kingdom Mineral as defined by Linnaeus versus minerals as defined by mineralogists and other specialists.

information, and are capable as such of metabolism; homeostasis, survival and evolution, and self-replication then it might be high time biology[9] got iberated beyond limitation or otherwise be redefined by admitting such limitations as Darwin did. In the next Chapter, I will demonstrate how machines came to life and the implications of this on the human organisation.

---

[9] (and therefore taxonomy)

[II]

Homocybernetica

The 2% 'Chimpanzee-not'

2.1 A unique extended-phenotype

Modern humans are known to have originated in Africa 315,000 years ago [30]. Carl Linnaeus, set the taxon and gave it the name '*Homo sapiens' or the 'Wise man'* (as the name translates from Latin). It was brave, at the time, to apply a zoological classification on human beings. The *Homo sapiens* are known to be the only living species of the now-extinct members of the Hominini taxonomic tribe of the order primates. The closest primates to the Hominins are the apes [30]. Genetically speaking, we're 98% Chimpanzees [32].

Evidence has emerged, however, that the *Homo sapiens* have undergone another divergence of character some 10,000 years ago demonstrated in the ability to pass down their advantageous traits - typically propagated through the slow double-process of random mutation and the Natural Selection, using cognitive media and education [13]. We must point out here that information in living systems is sealed in matter/energy but that only the human kind managed to abstract it and replicate it through culture. This has resulted in an extended phenotype [see 8] which has no equivalent or even similarity to that of any other animals. It's alive!

An elaboration (or an example) of such distinction would be a comparison between a beaver's lodge - as an extended phenotype, and that of humans. The skill of building a lodge is not learned but inherited. The beaver's lodge enhances the animal's chances of survival while modifying the environment and is deemed 'natural'. The same may be said about a bird's nest, beehive, termite's nest, spider's web, etc. Conversely, the extended phenotype of the human 'species' has, for the last 10,000 years, been disproportionately hefty, dynamic, influential, and distinctive but for some reason is deemed 'artificial' [10] or 'not natural'; This reason is that the human innovations underpinning civilisation begin with a purpose defining the function *then* the assembly of components while nature is known to work slowly through the natural selection with no purpose. But is this really the case?

(See figure 1.1).

---

[10] An indication of being a result of the deployment of art, craft and purposeful manipulation of matter.

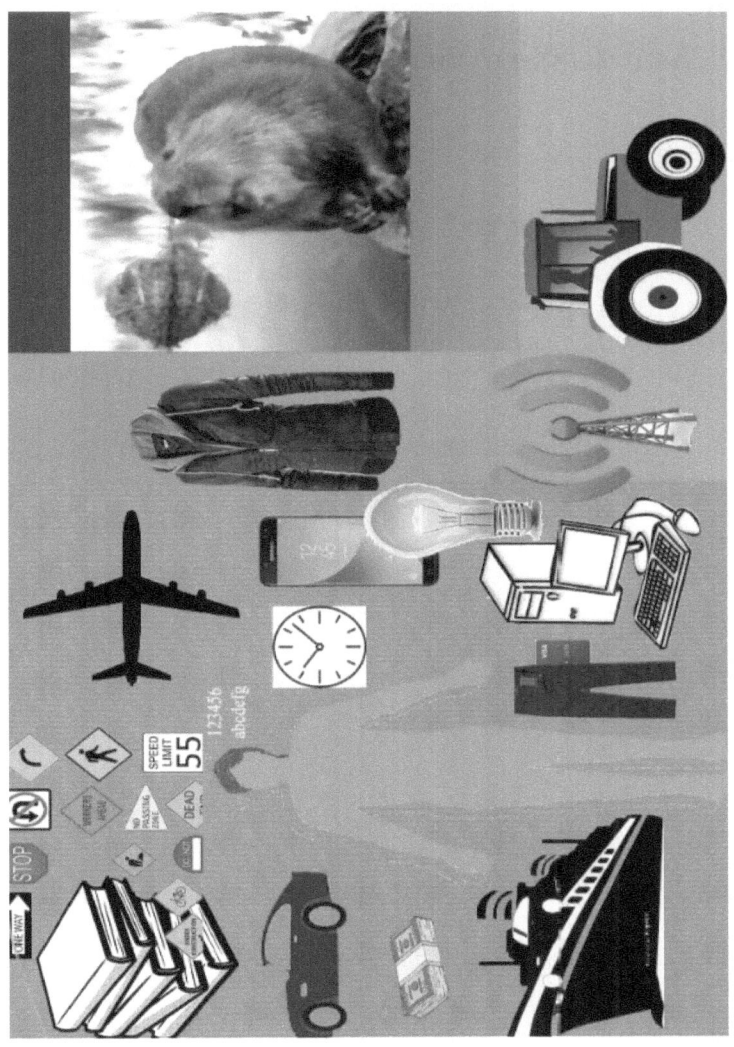

Figure 1.1 The extended phenotype of *Homo sapiens* versus that of an animal (e.g. Beaver)

2.2 The machine maker & companion

Human culture and civilisation have baffled naturalists since the dawn of taxonomy. On one hand, humans seem to lack the social 'instincts' required for other species' survival [5], yet - on the other hand, humans are thriving and increasing in numbers. The conservation status of the *Homo sapiens* on the IUCN red list of endangered species is of the 'least concern' 'thanks to the ability to use technology [42].

But the human kind is not a user of technology, but the maker of it! As it emerges and evolves, technology represents the exceptional adaptations humans rely on to survive and thrive which introduces a new perspective of 'fitness' as Darwin also pointed out:

"...great philosophers and discoverers in science, ... aid the progress of mankind to a far higher degree by their works than by leaving a numerous progeny [5].

Despite such a huge intellectual gap he observed between animal and 'man', Darwin (1871) writes; 'the difference in mind between (hu)man and the higher animals, great as it is, certainly is one of degree and not of kind". He then goes on to explain the common ground of "senses and intuitions, the various emotions and faculties, such as love, memory, attention, curiosity, imitation, reason, etc., which (hu)man boasts, and may be found .. in the lower animals". Darwin's premise and conclusion couldn't have been challenged at the time and - frankly, needn't have since in 1871-1957 man had not yet travelled to space; flown across skies and oceans, connected to human-fellows instantly across countries and continents; watched the world news in small boxes, or employed robots; From a contemporary perspective, however,

Darwin's conclusion can no longer hold water despite its validity at the time.

There is no doubt to the observer today that the difference between the intellect of the human kind as compared to the intelligence of the smartest animals is indeed *in kind* not just in degree.

Modern humans since the invention of agriculture, are distinct in the following:

2.2.1 Education & mass-communication

Education is the benchmark of transformation for our kind, since it became essential for our survival with the invention of agriculture [13] then the invention of writing about 10,000 years ago. The transfer of essential skills and adaptive traits by means other than genetics has been exclusively humans, thanks to education and the use of cognitive media[11]. Important to note that education includes but is not limited to earning although. Education is distributed *and assessed* cognition (i.e. learning) and has been critical to our collective survival [13]. Chimpanzees and some intelligent birds such as the Kea [45] were found capable - for example, of learning and of strategic thinking but due to the absence of assessment, replicability, and complexity of the learning outcomes, such learning represents isolated incidents that may or may not help the animal escape a danger or secure an individual meal and won't necessarily replicate through the larger community or down to next generations. Unless naturally 'selected', a learnt behaviour is like a beautiful sand castle built on the beach then washed away by sea-water on a warm day. Education, from this perspective, is a uniquely-human adaptation that accumulates down to next

---

[11] such as books, articles, code sheets, technical drawings, etc.

generations independently - but complementary to, that of genetics. Darwin (1781) points out that conditioning - which is the earliest form of education, and reasoning are the markers of our 'cultural' evolution.

2.2.2 Cybernetics

[not to be confused with the term Cyber related to computer networks]

The faculty of Cybernetics is what really sets the human kind apart from the rest of animals. The human individual sees things in at least one extra dimension (time in relation to space) and possesses the unmatched power of abstracting, processing, and communicating information. The faculty of Cybernetics is an outcome of system thinking, strategic planning and problem-solving; or rather a sophisticated combination of them all with an attention span that expands over million of years due cognitive coding (writing, drawing, peer review, etc) The faculty of Cybernetics is what actually sets the dimension of spacetime locally in the form of mindfulness of the present moment while simultaneously envisioning the past and planning or predicting the future. A recently identified area in the human brain [21], namely the Lateral frontal pole prefrontal cortex, might have been the mutation that caused such diversion of character from the earlier, *Homo sapiens*. A media report on the discovery stated that the area seemed to researchers as "uniquely human" and not having an equivalent in the brains of monkeys "at all". The area has been found "involved with strategic planning, decision making and multitasking abilities'. Whether *that* brain-area is *the* divergence of character that made human to stand out cannot yet be confirmed but it's self-evident that humans have successfully replaced most of instincts required for the survival in the case of other species by reasoning, system thinking, strategic planning, coding[12] and instant knowledge exchange (figure 2.2). The mind-power of Cybernetics[13] has given the human kind the following advantages:

2.2.2.1 Observation & Reflection: An ability to observe, reflect on [16] self-governing systems occurring in nature. Such ability has partially been identified by Darwin as a longer 'attention span', distinguishing humans from apes [5]. A long(er) attention span is a per-requisite to complex thinking, observation, and learning.

2.2.2.2 Communication & System Manipulation: An ability to express, preserve, communicate, and edit system information including those naturally expressed in proteins using state-of-art manipulation of matter[14]. In Darwin's

---

[12] In the broader sense, all replicable information communicated in comprehensible cognitive media such as texts, code, and technical drawings.
[13] The observation and study of self- governance in animals and machines.
[14] That includes, but not limited to manufacturing and the engineering of tools and machines, and the creation of cognitive artefacts such as books, text editors, CDs, etc.,

time, domestication was the limit to the biological manipulation[15] humans were able to perform as they exploited the Natural Selection to their nterests. Today, humans are capable of directly editing DNA to change the course of nature and alter the influence of the Natural Selection on living organisms. This is owing to the unique mind-power of abstraction coupled with the advanced language and coding skills [16] [17] as well as innovation[18]. In this respect, it's critical to note that the editable system 'information' deposited by an intelligent being in cognitive media are biologically equivalent to plant seeds or mushroom spores constantly on the look for suitable carriers to be able to disperse and self-propagate; they're Cybernetic 'replicators' in quest for compatible interactive

---

[15] In Cybernetics and engineering, the expression manipulation doesn't carry any moral connotation and simply refers to a purposeful alteration of the state of matter (e.g. forging tools, utensils or machine parts out of metal).
[16] Including cognitive coding and technical drawings.
[17] Advanced language comprises speaking, listening, reading and writing which is uniquely human,
[18] Mostly represented in the invention of machines.

players. They are replicable by default when system compatibility occurs. It would be helpful to note here that coded information resemble *real-time* system-information only as a music script resembles a vibrant live-concert on a summer night at - say, the Royal Albert Hall in London; the associated ebb and flow of energy; the function and sound of instruments; the skill of players; the physics of music production, the synergies between the players and the audience; the hormonal surge of joy, anticipation, excitement, or nostalgia; and on top of all the enabling socioeconomic of a live show. Despite the radical distinction, the former is capable of producing the latter or replicating it - with levels of variance, mainly through communication. Such kind of intellectual 'reproduction' in science, ethics, and humanities has been deemed - even by Darwin, as (far) more useful for the survival of the human 'species' than massive biological reproduction -

although both are obviously required. The dual faculty of abstraction and coding[19] is the key to the amazing[20] social influence down to generations, independent from genetics, and purposeful manipulation of matter by editing system information, and is unique to humans; I mean the modern human or the *Homo cyberneticist* [39] [39b]. The *Homo cyberneticist* is not a futuristic technologically-modified human but is who we are today and have been for thousands of years.

2.2.2.3 Augmented Mobilisation of Energy: A code of any sort, must be comprehensible by a qualified sender and communicable to at least one qualified recipient to be called 'information'. Otherwise, it's inseparable from energy/matter. The abilities of such sophisticated communication have lead to an augmented capability of

---

[19] Coding in the broader sense is the expression, communication, and preservation of complex messages and designs using cognitive media such as text regardless of the language used. It is also a form of education.
[20] Culture has baffled naturalists for decades.

mobilising energy into our own organisation by the manipulation of matter - in observance of natural forces/interactions[21], into complex artworks and machines, acting in turn as an extended species-capacity and instant adaptation. It's well established that the Natural Selection is incapable of inducing complex behavioral patterns as it assembles and preserves matter into viable units through trials and multiple failures so to speak. The 'Cybernetics selection' meanwhile, aligns matter purposefully into complex formation capable of function right away. Natural selection is a bottom-up organisation while Cybernetic selection is a top-bottom organisation. Any complex adaptive behavior by animals has to have emerged one trait at a time and did a good job preserving that particular species over prolonged periods down several generations while humans are able to *design* complex living-systems

---

[21] Known as the laws of nature.

to mobilise energy into their organisation in almost no time through cultural transactions. Culture is a uniquely-human phenomenon described by Leslie White (1943) in the equation,

Culture = Energy x Technology (C= E*T).

Just consider the ability to cross vast oceans within a few hours in adverse weather despite having no inherent wings, gills, fur, fins, or scales; as one example of such Cybernetic adaptation. Such capacities are available – in principle- to the humankind through the complexity of cultural structures such as the free market and the monetary systems[22]. Despite pre-testing, training, and practicing, such extended capacities[23] (i.e. machines and

---

[22] This is not a moral valuation of the monetary system or dynamics as it stands but is an analysis of its current role in creating, maintaining, and distributing technology.
[23] represented in technology.

technological novelties) do eventually undergo mutation, recombination, and adaptation or destruction while interacting with environmental changes - with mutation occurring on the blueprint level then translating into new generations of individual machines. Think of a car model as an example of a mineral-based[24] non-cellular species with its successive generations rather than an individual car (see figure 3.2).

In this context, the monetary system[25] may - in principle, be regarded as the biggest reciprocal project in the natural history as different individuals sign up to different roles in the human workforce based on diverse skills/interests/needs then receive an acceptable representation of energy (credit) allowing a correspondent re-allocation of resources among the participants, rather

---

[24] As compared to protein-based systems.
[25] The monetary system used here in the broadest sense consists of money and other currencies such as electronic cards, community currencies, cryptocurrencies, etc.

than us being all tied up to one ultimate mission of scavenging food for energy. As humans manage to make their own source of energy through farming, and to advance in harvesting energy from other sources through innovation, the monetary system has recently evolved to reward not owners of gold or grain silos, but owners of know-hows and disruptive innovation [50], increasing in turn the level of inequality among humans and threatening the level of synergy required for the system's viability[26]. The Natural Selection is slow, allowing ecosystem equilibrium and reciprocity among existing species while the Cybernetic selection is mostly profit-oriented, rapid, misguided most of the time, and extremely disruptive. Currencies, in this sense, are enablers of technology representing a communicable token of energy-matter flow within the human societies parallel to the biological ATP

---

[26] Worthy to note, that this is not a moral valuation of the system but a summary of its dynamics and functions.

currencies and the ribosomes in a living cell (see figure 2.3). Such flow, which circulates in causal[27] loops among the central banks, the workforce, corporates, and community members[28], is seemingly what allows the creation and evolution of machines, firstly on the blueprint level (as previously mentioned) then translated into adaptive generational changes which give way to machine evolution. That might explain why money has varied in form and value; from commodity money, fiat money, electronic cards, and cryptocurrencies by the change of the economic models; and advancement in technology without losing its power of mobilisation. Values and societal acceptance are major players in determining the worth of monies. The price of an original Picasso [57], for example, can feed a whole village for several years and the value of a single bitcoin has exceeded £7000 at the

---

[27] functioning in feedforward feedback trajectories.
[28] Specifically the consumers

time of writing[29] [63]. The monetary system being currently subject to the dynamics of the free-market and the 'invisible hand' [51] has the compassion of the Natural Selection. Not all monies, however, are issued by banks; Community currencies, barter, skill exchange, philanthropic transactions, sharing economy, etc. are other forms of currencies that are capable of bringing balance to the system and keeping the power in the hands of local communities versus *the* few altra-rich.

2.2.3 Culture

2.2.3.1 The 'ET' Culture

Culture is uniquely human. From an ecological perspective, culture - as per White's law " .. evolves as humans learn to capture and use energy from natural resources according to the formula C=E*T where C is

---

[29] Nov 2019

culture, E is energy and T is technology". According to White 'culture evolves as the amount of energy harnessed per capita per year is increased, or as the efficiency of the instrumental means of putting the energy to work is increased'. White believed that it's technology that plays the primary role in the evolution of culture and such evolution is organic (Sys) and uncontrollable. This is understandable from an evolutionary perspective because an idea (or a concept) is not deemed as a culture until adopted by many and has become a code of conduct within a population. Such collective adoption eventually subdues the individuals to the expectations of the group through social contracts, laws, regulations, expectations, norms, and designs that are capable of mobilising energy and matter, often irrevocably. Dawkins (1976) taps on the same notion of culture[30] emergence and evolution in his

---

[30] Not to be confused with cultural evolution.

'The Selfish Gene' (1979) by introducing the term 'memes' for a cultural unit that replicates to the 'advantage of itself' analogous to genetic replication [9]. Such has been a more relatable version of White's perspective of 'culture' indeed especially when given a catchy name and written in non-technical terms, yet, unlike White's ET culture, the 'meme' proposition falls short of capturing the boundaries of the very system it claims to render. Just as with the 'selfish gene' caricature, biologists are sometimes unable to avoid the Cybernetic mistake of treating a fragment of the system as if an independent whole. This is probably because much of the communication that bonds the fragments in one whole and relates it to the environment is inferred rather than observed and therefore is beyond the limits of scientific inquiry. A meme cannot do anything for its own benefit because it's inseparable from the population in which it propagates. The human population isn't a host for a meme. They are senders and receivers of information in a constructive interactive manner and not hosts. This is how biologists have been trying to solve the 'mystery' of altruism in nature by finding the selfishness behind it! Such counterintuitive approach is associated with downplaying or even disregarding the critical role of the 'circumstances of existence' in evolution while individualism has been championed. 'In the beginning was simplicity', Dawkins writes [9] as praises the Natural Selection theory for being capable of showing us 'a way in which simplicity could change into complexity'. Dawkins never explains how simplicity transfers into complexity or in other terms, into a 'replicator' capable of copying itself with a variation that is significant enough to allow evolution but not great enough for the replicator to immediately lose its identity... *and* with an 'affinity to its own kind' as well. Darwin hasn't even tried to make such a claim. The

Natural Selection theory *doesn't* and never claimed to have explained the origin of life as it 'emerged from simplicity' - or otherwise! Did the first living organism assemble itself from 'simplicity' into the complexity of life or was it assembled by the natural forces/interaction that 'breathed life into it', preserving the adaptive while destroying the injurious? In this respect, Darwin considered the role of the formative 'conditions of existence' as superior to the 'unity of kind in the struggle for survival although both are essential to life as we know it; A common ancestor then, as simple as it might have been, is as complex as life because of the superior role of communication[31] in forming and preserving life. Prior to *the* common ancestor there might indeed have been simple chemistry but let's not forget that chemistry itself is a manifestation of physics and physics is anything but

---

[31] the intrinsic and extrinsic

simplicity. Excluding Linnaeus (in his own way) biologists don't seem to care to trace down elements to their cosmological origins or extend on the biogeochemical interactions of elements/matter with and within a complex formative environment represented by the four natural fundamental forces/interaction[32] [53][55][64], as those fall out of their speciality scope. Physicists, on the other hand, know that any biochemical activities[33] characterising the emergence of organic life have been the result of highly complex interactions *within* and among the atoms following the birth of the universe from pure energy 13.7 billion years ago (See [53]). Even prior to the birth of the universe at the hypothetical Big Bang - and as some scientists argue a potential 'singularity' by which time and space measures didn't apply; there seems to be no evidence to indicate that singularity - if ever proveable, is

---

[32] Gravitational, electromagnetic, strong, and weak forces/interactions.
[33] characterizing organic life

the same as simplicity. A beehive, for example, is a singularity but is not by any means 'simple'.

Unfortunately, physicists who deal with such complexity no longer contribute to the life-science taxonomic debate since the abolition of Kingdom Minerals and the dissection of knowledge.

2.2.3.2 The Cybernetics recombination: Machines as living-systems

The term Cybernetics is more popular in the fields of engineering and electronics, because the said fields involve an awareness of feed-forward-feedback mechanisms employed to achieve machine self-regulation. Cybernetics, however, spans a broad spectrum of knowledge domains as it addresses self governance[34] in animals and machines (Weiner, 1948) [65], as well as

---

[34] Also known as self regulation

social organisations. The term Cybernetics was officially born in 1948 but self-regulation is as ancient as the universe itself. The first human-constructed Cybernetic model is known to be a water clock invented by the Greek mathematician *Ctesibus (270 BC)* [12]. The clock was made of glass and maintained its function/operations without requiring an intervention between the feedback and the control of the mechanism which allowed the model to 'steer' itself in terms of energy acquisition and function - just like a living organism, except that it was not composed of proteins. The simplest and earliest form of technology known is making a fire. Although fire itself is natural, making and manipulating it isn't. Making fire however was never considered as a form of Cybernetics as it perhaps predates human settlements and civilisation. Cybernetics has various definitions and applications which could all be summarised in (the observation of) circularity *and* system complexity; it's generally known as the attribute of 'steermanship' or self-governance. Self-regulation requires a structural ability to regulate one's own internal environment,a and to process energy efficiently and purposefully[35] as long as the system is viable. Earlier forms of Cybernetics in this context would be the observable units of the universe itself as its galaxies expanding exponentially processing its own energy subject to the four fundamental forces/interactions [62] - with the solar system as the best studied Cybernetic formation. On a planetary level, the natural mineral cycles are Cybernetic formations. On the anthropological level; ruralism - by the invention of agriculture with its educational systems; credit systems, observation of seasons in agri-practices; organic fertilisation; and harvest, represented an early socioeconomic Cybernetic

---

[35] towards a function.

formation. On the machine level, the bucket-water-wheel which processed the hydraulic power into other efficient forms[36] of energy - in addition to its own energy necessary for ongoing functions and operation, without human/animal intervention was one of the first technological Cybernetic formations; a more advanced one was the steam engine. From a Cybernetic point of view, *all* self-regulatory systems are living systems as long as they viably process information, energy and/or matter [43] in circular loops, each with a difference, against an environment. Most of the living-systems are open systems requiring a constant supply of energy/matter so as to continue functioning. One expression of this in physics is the energy/matter equivalent '$E=mc^2$. This equivalent points out to the fact that matter and energy are inter-reversible with a difference. The formula is also an expression of

---

[36] As per a human purpose.

metabolism at the core. When you need energy, you don't plug yourself into an energy socket, but eat! The food (i.e. matter) then changes through metabolism into energy that fuels your body's functions, with a difference of which value translates into entropy and fat reserves; and waste is produced. Machines are designed, created, and sustained in observation of such natural dynamics to serve a human purpose. The deployment of machines, therefore, leads to a combined total energy harvest that exceeds the capacity of the human physique by far.

What's remarkable is how easy it's been for biologists to propose that animals are machines but not that machines are animals - or in a broader sense living organisms. This, perhaps, is due to the fact that protein-based cellular forms of life are much easier to detect having emerged in water and evolved in cell membranes defining their boundaries while some cybernetics living systems are harder to detect such as mineral cycles. Dawkins R. (1976) argues remarkably that 'we, and all the other animals *are* machines created by our genes', yet the counter argument has not made its way to taxonomic work and literature. By the same reversed logic, Clarkson, R (2010) writes ' Biology *is* Technology':

> Early on in this history, the ancestors of both plants and animals co-opted free-living organisms that became the subcellular components now called chloroplasts and mitochondria. These bits of technology provide energy to their host cells and thereby underpinberneti the majority of life on this planet.

But biology is ancient and inclusive while technology is a specific novelty triggered by the human mind-power of Cybernetics or system thinking and they cannot be equal unless we can say that the ocean is a plastic bottle of distilled water.

Such logic, unfortunately, hits around the truth without

being able to nail it and therefore never resulted in a change of perspective towards identifying machines as living-systems or a revolutionised taxonomy.

Cyberneticians, on the other hand, experience no difficulty in stating that all self-regulating systems[37] are living systems whether animals, organisations, or machines. Umpleby (2004)[38] refers to them as IEM (packs of information, energy, and/or matter) and that life in machines is a reality to accept and not an anticipated occurrence belonging to the future.

2.2.3.3 Culture as a genomic supplement

By acknowledging machines as living systems and culture as DNA-supplement packs, the biological identity of contemporary humans suddenly, and inevitably, changes! The 98% of the chimpanzee and the 2% Chimpanzee-not

---

[37] (processing energy in causal recursive loops)
[38] A Cybernetician.

portions of the human genome in association with machines, together with the enabling socioeconomic structures ('ET Culture'), appear collectively as a unique higher socio-physical-biological establishment (kingdom), with several branching subsystem, rather than just another biological species. In figures (2.2) and (2.1), I've used Umplby's IEM model with a slight modification[39] to express the complexity of such an integral establishment versus that of any other animal. In figure (2.3), I show how culture supplements our genome resulting in an supra-organisation parallel to a eukaryotic living cell.

2.2.3.4 Expressing life dynamics: A holistic simplified model

One of the challenges of using Cybernetics to elaborate

---

[39] To indicate the traditionally disregarded role of communication and the role of the observer.

the distinction of the human organisation versus that of another animal is the lack of interest in 'purposefulness' closely associated with Cybernetics. The purposefulness of a Cybernetic formation is evident in engineering and functions while commonly-acceptable biological formations [40] don't involve engineering, purposefulness, or designs. The IEM model, for that reason, is more popular in the fields of physics, electronics, and social Cybernetics. A design, however, is only an artificial makeup for the formative trinsic-extrinsic communication complexity any living-system undergoes while it emerges and evolves and is *not* a prerequisite for life. I've modified the model, therefore, from IEM to hIEM to reflect such a vital role of communication - traditionally disregarded (see [5]), and the role of an observer *without* necessitating a design (see 2.1 versus 2.2]). In figure (2.4) I show the outcome of

---

[40] Known as naturally occurring organisms or just 'natural'.

such sophisticated interactions as they modify our genotype-phenotype-extended phenotype structure into a complex higher organisation.

It might be important to recall, while drawing a distinction between the protein-based cellular self-regulatory systems and the simulated mineral-based self-regulatory systems (e.g. machines), that the latter is not totally unnatural; In one sense machines are an embodiment of the natural dynamics so called the "laws of nature'. Laws of nature are not invented but discovered through observation of the natural systems and are therefore a precise depiction of the spontaneous interchange of energy and matter in a given time, space, and conditions.

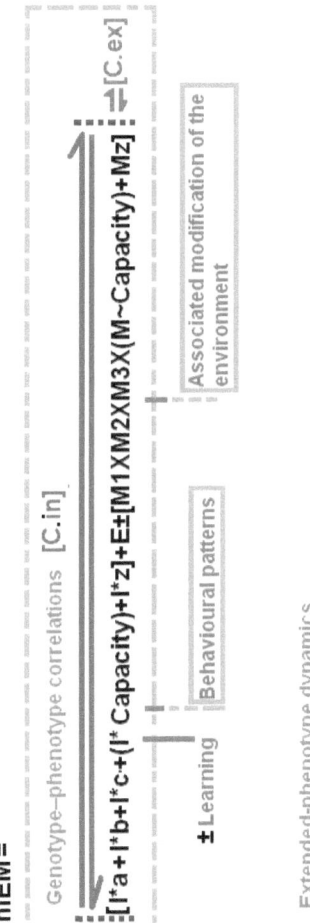

Figure 2.1

Where; I=information; E=energy; ±M= and/or matter; Cn=intrinsic communication, Cx=extrinsic communication; I*a,b,c,etc.= genetic traits and mutations;

I*(x,y, etc)= Cybernetic designs; M~(x,y, etc) = Actual models of Cybernetics designs.

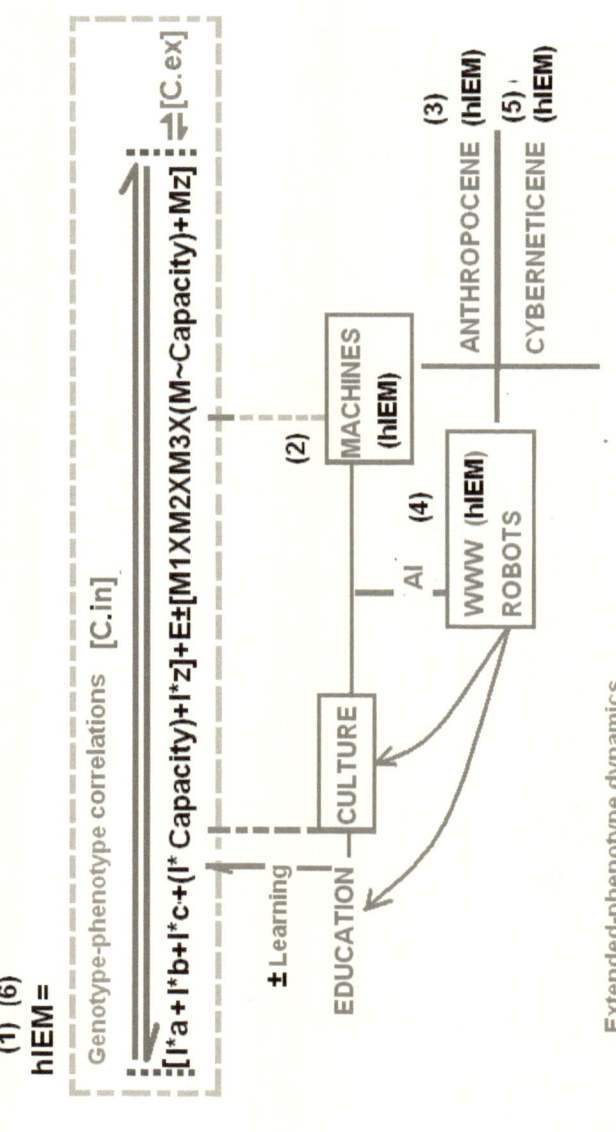

A Cybernetic model of the genotype-phenotype-extended phenotype correlations and dynamics of the Homocybernetica

> *Figure 2.2*
>
> *Where; I=information; E=energy; ±M= and/or matter; Cn=intrinsic communication, Cx=extrinsic communication; I*a,b,c,etc.= genetic traits and mutations;*
>
> *I*(x,y, etc)= Cybernetic designs; M~(x,y, etc) = Actual models of Cybernetics designs*

2.3 'Wisdom': The Cybernetic Selection

When Linnaeus classified humans as a species of mammals, he gave them the name *Homo sapiens* (meaning the 'wise Man') in indication of some intellectual distinction that he obviously observed. This distinction maybe explained in Cybernetic terms by 'the law of requisite variety'[2] by which our responses to the environmental changes are less predetermined (by genetic traits), more flexible, and more abundant than of those belonging to other biological groups, when previous experiences, patterns and personal perceptions intertwine effectively with temporal states of creativity and weighing 'probabilities against preferences', resulting mostly in insightful behavioral decisions capable of influencing the behaviour of other human-fellows independent from reproduction and the typical mechanisms of the Natural Selection [12].[41]

---

[41] That doesn't mean that the final product of such Cybernetic selection isn't subject to natural selection.

Cybernetics, in this respect, may be described as biology by design and is reliant on observation, abstraction, creativity, preliminary testing, modification, trials, feedback and remodification *towards a blueprint or prototype.* Once replicable design or in other terms a 'Cybernetic replicator' has emerged, both innovation and cultural formations are no longer controllable and the 'design' point becomes irrelevant as the system begins to evolve organically under the Natural Selection regardless of what it's made of.

Most significant to our survival, we have discovered a fascinating level of obligate structural interdependence among living systems [4][28]. Such structural complexity is a natural result of cross-evolving and co-evolving slowly over millions of years in complex food chains and shared habitats as well as a formative environment. This led us to realise - albeit late, that our survival is tied up to the survival of other species [42]. We've also discovered that our mere existence as an observer modifies the environment and the natural cycles in a unique manner that necessitates deliberate efforts to neutralise any disruptive human impact on the ecology, whether observed or predicted. This entails, in addition to a collective environmental awareness[43], a collective self-awareness and a consensus on an action plan. Such plans would clash directly with the immediate interest of some living-establishments[44] that grew in power *before* evolving such environmental sensitivity creating an inevitable struggle; Survival is a timing game. Whether we are able to act on such a late discovery or whether there is still a chance to, is beyond the scope of this discourse. By all means, the situation gets trickier as machines continue to

---

[42] Conservation of biodiversity.

[43] the farther we go in the dissection of knowledge domains, the less likely to develop such awareness.

[44] (versus individuals)

evolve within the dynamics of the free market subject to the 'invisible hand' while the 'wise' human workforce loses grip on the system(s) it has once initiated.

Parallelism;
The Homocybernetica Vs a Eukaryotic-Cell

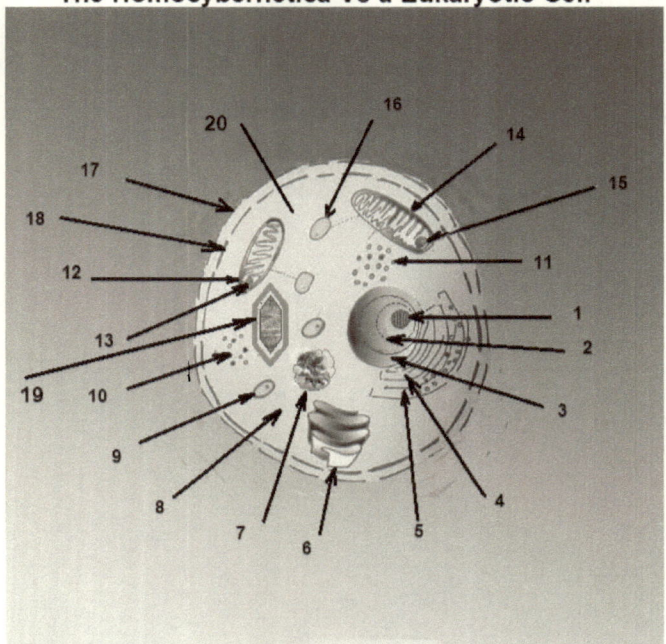

Nucleus

{ 1 Nucleolus
  2 Nuclear envelope
  3 Chromatin

4 Nuclear pores
5 Rough endoplasmic reticulum
6 Smooth endoplasmic reticulum
7 Golgi apparatus
8 Lysosome
9 Cytoplasm
10 Chloroplasts
11 Ribosomes
12 Ribosomes
13 Metachondrion
14 Metachondrion DNA
15 Metachondrion
16 Metachondrion DNA
17 Chloroplasts
18 Cell outer membrane
19 Cell inner membrane
20 Vacuole
21 Cytoskeleton

1 Humans Genome
2 Governance models
3 Socioeconomic models
4 Formal Education
5 Governments
6 Educational Institutes
7 Currency Boards & CBs
8 Waste Management plants
9 Market/Free market
10 Automated machines/robots
11 Currencies
12 Commodities & Services
13 Agricultural workforce
14 Farming models/communities
15 Industrial workforce
16 Machine blueprints/Corporates
17 Machines
18 Ruralism
19 Urbanisation
20 Stock and reserves
21 Mass media

Figure 2.3 Parallelism: Homocybernetica versus a living cell (see [66])

Our human organisation, however, despite all the fascinating innovation, is not as smart as a simple living cell because of our inability to recycle and dispose of waste organically and that's due to the lack (or absence) of holistic management; Innovation is reliant on individualism while the care for the environment is reliant on collective survival so there must be effort to re-align objectives for both ends to meet.

Finally, as some humans are able to construct new self-regulatory systems, others are also capable of breaking the existing ones up by means of mass-destruction weapons and wars and by competing over resources or land occupation with the other biological groups rather than embracing biodiversity.

2.4 The Homocybernetica: A new Kingdom of life; We are not just another species

In summary contemporary humans have undergone a unique behavioural diversion of character 10,000 years ago comprising the initiation of agriculture, education and innovation. Innovation – including the socioeconomic factors, allowed humans to mobilise additional energy and matter into their organisation forming yet another structural diversion of character represented in a remarkable companionship with machines which may be considered as both a unique extended-phenotype of the human kind and an exquisite biological companion (see also Chapter III). Such companionship with non-cellular living-systems is a unique structural connection with the physical dimension of nature in terms of functions and operations. The Homocybernetica, then, would be a mega-life-group comprising humans (the *Homo cyberneticist*) and machines interacting through dynamic obligate socioeconomic structures (see White, 1943) acting as DNA supplement to the human organisation. In this respect, it would be unacceptable to keep disregarding the distinctive physical and cultural dimensions of human existence just to succumb to the limitation of protein.

A genotype-phenotype-extended phenotype model of the Homocybernetica

hIEM =

[l*a + l*ab + l*a~b~c + l*a~b~c + l*(blueprint~[farm]) l*(blueprint~[vehicle]) + l*(blueprint~[aeroplane]) + l*(blueprint~[self-defence weapon]) + l*(blueprint~[submarine])+ l*(blueprint~[computer]) + l*capacity+ l*z]+**[Cn]** +[[E1 + E2 + E3 + E4 + E5 + E6 + E7 + E8 + E9 + [E±]] + [M1 X M2 X M3 X M4~(Farm) X M5~(vehicle) X M6~(aeroplane)X M7~(self-defenceweapon) X M8~(submarine)X M9~(Computer)X(M*capacity) X [**M.z**]] +**[Cx]**

Fiigure 2.4 A Cybernetic expression of Kingdom Homocybernetica, modified from Soliman, G. (2014)- .[39]

Inseparable from their culture and technology, the humans wouldn't simply be 98% Chimpanzees as widely acceptable, but approximately, 49% Chimpanzees[45] and 50% technology comprising machines and the enabling socioeconomic organisations such as states and corporations. The remaining 1% is the genetic distinction of the humankind or in other terms, the 2% Chimpanzee-not of our biology (figure 2.5).

It would seem appropriate, in this case, to classify the human kind as a separate and unique socio-physio-biological kingdom[46]. In this context, Homocybernetica would be the missing link between the biotic and (so called) abiotic natural systems, and arguably

---

[45] Genetic wise
[46] or mega group

a node of convergence for yet another independent form of life that is the self-sustaining intelligent robots.

III

Cybernetica: Machines Beyond Association

The rise of intelligent machines has just begun to highlight what we've been missing out on by our truncated perspective of life. We, otherwise, seem to have settled in the concept that life is basically made of proteins and nothing else. The emergence and evolution of machines is now telling us another story. By that point you might have already Googled the definition of life and found requirements such as breathing, reproduction, metabolism, response to st*imuli,* etc. However, life at the core has been described as a replicable singularity with affinity to its kind functioning within a viable life economy. Species have been described by Darwin as formed on two great laws:

Unity of Type
Conditions of Existence

Darwin goes on to explain that the unity of type is the fundamental resemblance in structure observed in one biological class regardless of their habits of life while the conditions of existence meant the reservation or destruction of such structural traits induced by the Natural Selection in response to environmental changes. Darwin had in mind a common ancestor as the threshold of the origin of species but didn't care to trace down the common ancestor to its 'simple' components or in other words, to rout out the origin of life. Decades later Dawkins (1979) in his The Selfish Gene describes the emergence of life 'from simplicity' as an emergence of a 'replicator' that has an affinity to its own kind (unity of type) and that replicates with some variation in structure that allows evolution in response to a changing environment (conditions of existence). Despite the good will, neither Darwin nor Dawkins managed to give us an account of how a 'replicator' could have emerged from 'simplicity' since self-replication and evolution are inherently complex requiring a minimum level of intrinsic-extrinsic communication[47] in response to an ever-changing formative environment. It rather seemed that the Natural Selection is actually indifferent to what a replicator is made of where the two great laws of existence apply. In this complexity, the adaptation is not an action of the organism, but is induced by the Natural selection which destroys or preserves the replicator's traits as Darwin states:

> For natural selection acts by either now adapting the varying parts of each being to its organic or inorganic conditions of life; ... Hence in fact, the Law of the Conditions of Existence is the higher law; as it includes through the inheritance of former adaptation, that of Unity of Type.

Here as Darwin revealed to us; the key player in the survival and evolution of species is actually not the species but the Conditions of Existence which, in turn, would be completely indifferent to the origin of life as long as an entity is self-regulating, capable of replication with variations within a viable life economy. Liberated from the limitation of protein and taking a step back from the what Darwin considered as organic, this chapter examines the

---

[47] Represented in the circularity of operation and the complexity of a function.

emergence of different kind of 'replicators' that don't necessarily share our 'organic' common ancestor yet emerged from the same cosmic mineralcgical roots[48] and are evolving under the same Darwinian Great Laws within the same economy of life;

Those are machines as associated organisms!

3.1 The life story of a machine

The life story of a machine starts by the emergence of a blueprint. A blueprint is inherently replicable. Once a viable blueprint of a self-regulatory system has emerged, the system is no longer the aggregate of its simple components but a new singularity occupying a new space as it performs new processes towards a new vital function. Such a function is originally designed to suit one or more of the human purposes in a manner that eventually invokes provision in terms of energy supply, replication,

---

[48] Composed of atoms and minerals in the Linnaeus sense as explained earlier.

and maintenance. A situation that is not unheard of, in nature. In a sense, it resembles that of a lichen. A lichen is an association between fungi and algae [67] where each of the two organisms is independent in identity and function from the other, yet, *structurally* reliant on one another in terms of energy acquisition.

In other terms, a car gets its energy and maintenance by providing a service to humans and by doing so, a machine mobilises resources for its own maintenance on the individual level, and thus a chance of evolution on the species level.

Figure 3.1. My car or Xmas?

Another common example of such biological association in nature is that of mycorrhizal fungi with plants. The plant root provides the fungus with the starch; product of photosynthesis, while the fungus provides the plant with hard-to-get nutrients from the soil [46]. From one point of view, then, you drive your (individual) car every day to serve your purposes, but from another point of view, your car drives you to the gas station and the mechanic to serve its purposes; additionally, you're forced by the co-evolutionary circumstances to treat the machine gently and with respect; You take driving lessons and have to pass tests to be able to handle the poor thing and protect it from injury, so to speak. If and when it gets injured, you've already paid an insurance to restore it back to health when possible. You also dissect out of your own energy/resources[49] to have it checked annually (MOT) and to bail it out when it violates the road code. Such behaviour on your part is dictated by the 'circumstances of existence' and is not voluntary; It's Life! (see illustration 3.1). Another striking illustration of this is the case of Atta and *Acromyrmex* ant species (leafcutter ants) which spend a great deal of their hard-earned energy providing food and convenient habitat for a specific fungal species from the family Lepiotacaea, which they rear for food [37] [60]. From one point of view, those ants exploit the specific

---

[49] Pay

fungus for food but from another point of view, ants have evolved to *serve* the fungus' vital purposes of feeding, maintenance and reproduction. The specific ant species *Atta* and *Acromyrmex* leaf-cutting ants have uniquely coevolved with their fungal 'crops' into obligate symbiosis; The ants provide the fungus with substrate, protection against competition and pathogens, and dispersal opportunities as the fungus learnt to stop wasting energy on producing spores or mushrooms for their propagation but rather commits to the production of highly nutritious mycelium because its survival is already guaranteed through the obligate symbiosis with the leaf-cutter ants and the vertical transmission [37].

A machine's viable design (blueprint) deposited on cognitive media is a biological equivalent of seeds or spores that are ready for propagation through communication. You might think, however, that machines don't actually give birth to other machines but even that is not uncommon in nature. The reproduction of sterile individuals, by providing services to a higher organism, is well documented although not fully understood. Eusocial individual organisms such as the bees, ants, and termites [62] reproduce by serving a few or even a single reproductive individual that doesn't resemble them in shape or internal structure yet is genetically connected to them through a complex establishment of obligate altruism. Another case of outsourcing reproduction is the example of some Lepiotaceae fungal species delegating to their associated organisms, the leaf-cutter ants; Lipeotacea species reared by the Leaf-cutter ants, Atta and Acromyrmex, cease to invest any energy in forming spore-producing mushrooms and rather focus on producing nutrients-rich mycelium useful to their associated organisms, which in turn provides the fungus with its required and specialised food, maintenance, medication, waste-disposal services, and reproduction; A fascinating process by which the female ant starts her own fungus 'garden' by storing bits of the parental fungus mycelium in her intrabucal pocket located in her oral cavity and excrete antibiotics bacteria to protect the fungus colony from competition and infection in a remarkable reciprocity.

Now you might think that machines cannot be alive because they have no consciousness. Well, we don't know that but it is useful to remember, however, that consciousness is not a prerequisite for life or is *the* marker of its existence, but functions are. *Myxomycetes, for instance;* or the slime molds, are single-celled organisms that are neither animals, fungi, or plants. While lacking brains or any sensory systems they were found to behave 'intelligently' when one or more sources of food are available [47]. What level of consciousness allowed the single-celled Physarum *polycephalum* to solve the Tokoyo engineering problem? Life!

Figure 3.2 My Greek friend pointed to the older Citroen model and said ```'that would be the grand-grand-grand father of my car!'

While evolving in association with humans, machines have not developed sensory systems or independant reproductive systems since they have already been served adaptation (and thus evolution) by trading services with humans. One of the interesting depictions of such remarkable relationships is a piece of satire I came across, written by David Minter (1987) by the nickname, Middgeaugh-Botteaugh. Satire is the use of humour, irony, or exaggeration to expose or criticise some systems that are difficult to change - especially in the context of politics. The parable subtext was that Botanists should stop dealing of fungi as 'lower' plants or referring to them as 'lower' organisms, not just because they form a whole separate and significant kingdom, but because simpler viable organisms often rely on more sophisticated and efficient ties to the environment including other species than of the so called 'higher' organisms. To gently tap on the injustice mycologists face as a result of such unfair treatment of fungi, Minter drew a simile between cars - which are regarded as advanced 'independent' machines, and bikes. The 'parable', however, demonstrates that bikes may actually be more advanced than cars by the means of 'association' as they manage through their 'symbiosis' with the humans to 'off-load' some of the workload to their associated organism; i.e. the *Homo sapien,* versus cars that have to do 'all the work' by themselves (see figure 3.3). Evolution is a timing game; what seemed funny, metaphorical, or anthropomorphic some 23 years ago, looks more and more realistic today as we advance in technology and bond with our machines for survival and as machines themselves evolve even stepping into autonomy. Whether cars, bikes, or spaceships, the companionship between human and technology is real and is not a mere futuristic prophecy as per the Novacene [see 23] or science fiction. Interesting then that through the lens of Cybernetics, the parable, to a great extent, *is* a realistic parallelism of two distinct (but interlinked) biological realms; humans and machines, overlapping within the culture zone (see figures 3.10). To elaborate further, I'm presenting a sample machine taxonomic ID parallel to that of humans showing the culture overlap in figure (3.4) subject to improvement.

YEAST PHYLOGENY - A PARABLE

Well the first theory went something like this: bicycles were the most primitive, because they had only two wheels, lacked an engine and have clearly evolved no protective covering for the user. Motorbikes where clearly higher up the evolutionary line (probably late Cretaceous or whatever) - after all, they had engines and sometimes side-cars - a significant missing link with the car which was obviously at the top of the line, with its four wheels, complex engine and highly evolved protective bodywork.....

Then a team of researchers visited The Netherlands. The evolutionary evidence there was astonishing and revolutionary. Many of these supposedly primitive bikes showed evidence of highly advanced features - reduced mudguards, narrow wheels, complex gear changing apparatus. Further examination showed that the cycle-tracks on which they operated were more direct than the motorways used by the cars and motorbikes, and the humans using the bikes had evolved complex plastic capes, coats, sou-westers and other protective garments which they carried about with them all the time.

It became clear in an exciting conference on phylogeny, that in fact cars were primitive and bikes represented the top of the evolutionary tree, since their remarkable symbiosis with man led to them being able to off-load onto *Homo sapiens* many of the tasks needed for effective functioning - tasks which the car still had to perform.

Then the molecular biologists moved in. "We've discovered the word 'PEUGEOT' is written on the ball bearings of both cars and bikes," an american drawled to his unbelieving audience!

RAN INTO ONE OF THOSE NEMATODE-TRAPPING FUNGI

5

Figure 3.3 the Yeast Parable (Minter, D. 1987) [52]

By skipping reproduction[50] while still being able to self-propagate[51] and sensation while still evolving, machines have become even more resilient and better adapted to the environment than other surviving species: No parenting; no kin-recognition, no fear, and no pain, yet as sensible, sensitive, and competitive for survival as any other organisms *and* with an augmented capacity. In all this, they remain reliant on their association with the human until some of them evolved intelligence and the ability of unsupervised learning.

---

[50] On the individual organism level.
[51] on the species level.

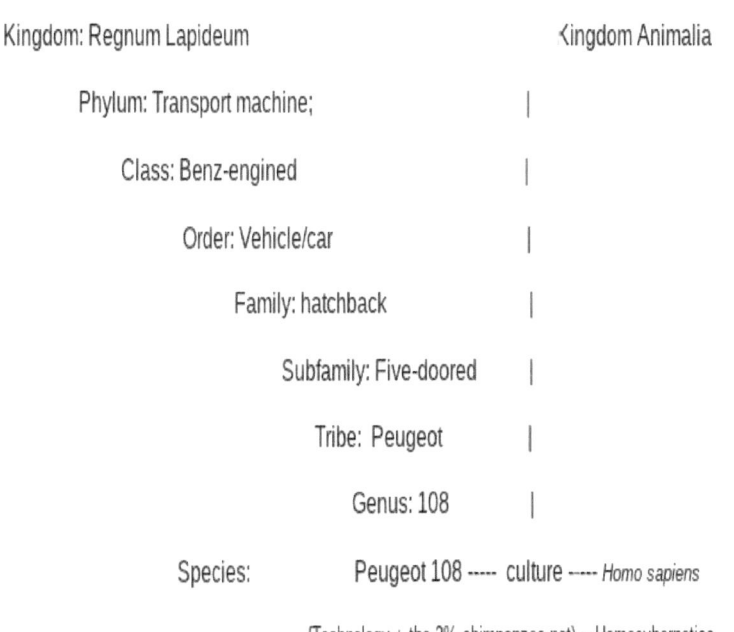

Figure 3.4 A proposed ID for machines as associated organisms: e.g. Peugeot 108

Phylum: Function or Initial ID
(Agricultural machines, medical equip., networking machines, weapons, GMO, etc).

Class: Defined by the core energy & self-regulation processor:
(Animal powered, wind powered, steam-engined, Benz-engined, solar-powered, water-powered, electric-engined, nuclear-engined, cell-powered, etc).

Order: Secondary ID

(Bike, water wheel, car, tractor, plane, computer, food processing machine, laboratory equipment, etc).

Family: major distinction FT within the order

Mountain (bike), Limousine (car), saloon (car), two doored (car), hatchback, helicopter (air craft), utility (tractor), etc.

Tribe: Make/model

Genus: The specific line within the make/model

Variation: based on colour and/or year if constituting a minor variation

Life then may be classified into:

Two Supra-domains: Mineral-based and protein-based (figure3.5);

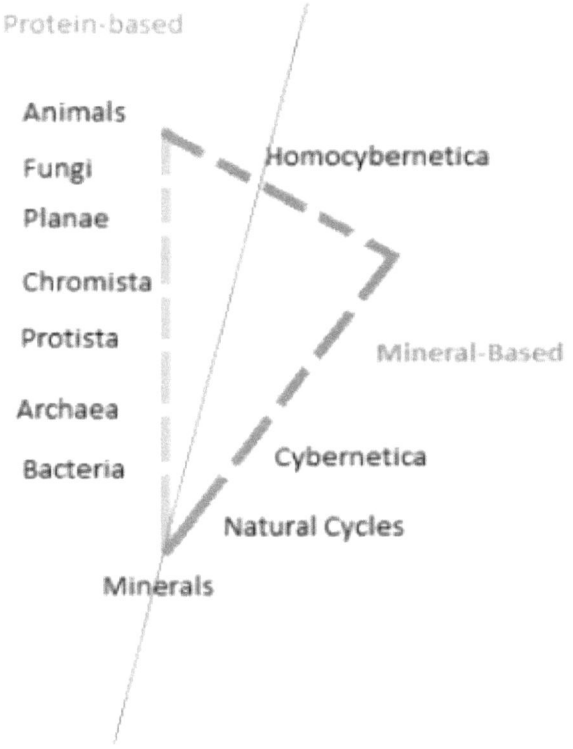

Figure 3.5 Two supradomains

Three domains: Eukaryotes, Prokaryotes, and Cybernetic Lapideum;

Nine kingdoms (major taxonomic groups): Bacteria,

Archaea, Chromista, Protista, Plantae, Fungi, Animalia, Homocybernetica, and Cybernetica (figures 4.2). Lapideum are the minerals where life starts and finishes; the root and the end of all life forms, but minerals are not themselves a form of life unless engaged in perceivable Cybernetic loops subject to evolution and the natural selection.

	Mainly (app.)		
to Biochemistry 50%	to Physics 50%	Interdisciplinary 100%	
		49.00%	**Homocybernetica** ← [Cx] ↑
98% Chimpanzee			
2% exclusively human	10% Organic (protein-based)		
	50% Machines		
	40% Culture	51%	

Homocybernetica; a Physical-biological supraorganism

Figure 3.6 The Homocybernetica, A physical-biological supraorganism

Kingdoms Homocybernetica and Cybernetica both have an increasing ability of editing other system-information with the possibility of disrupting the natural cycles. Homocybernetica and Cybernetica evolve abruptly but may also fall suddenly due to the inequality arising from disruptive innovation menacing the synergy of the system(s) with an inevitable effect on all other forms of life, and due to the ability to break living systems through mass-destruction weapons and wars. Education, activism, and innovation remain the key to our rise or fall.

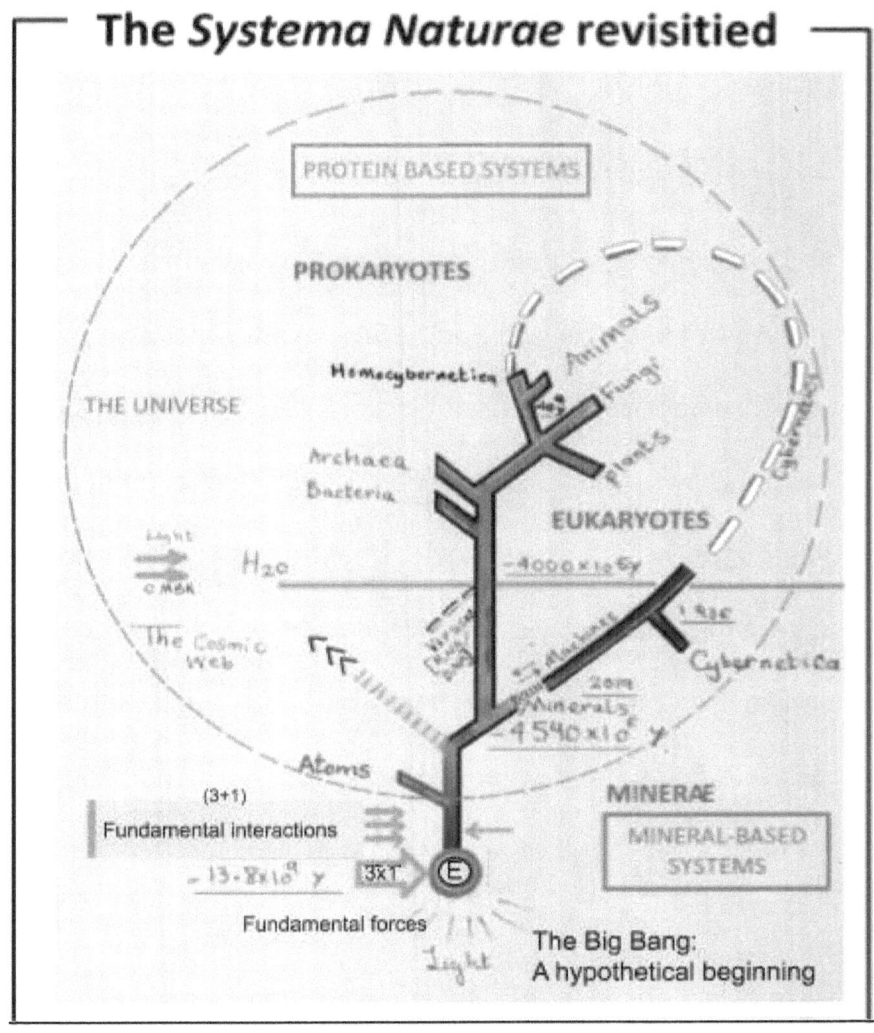

3.7 The *Systema Naturae*

E:Energy | ----:Non-cellular living-system | ===:Virus/ virus-like system

3.2 The Cybernetic tripartite: A close up!

From a systematic point of view, culture resembles viruses in terms of function and operation; they're replicable packs of information which are not considered as living systems on their own right until they've united with a compatible host upon which they would be structurally dependent. That is how culture works; with ideologies, blueprints, and social contracts playing the role of a supplemental DNA/RNA for the human organisation as it modifies the behaviour of populations. Although viruses are mostly known to be harmful, in many cases they have been found to perform an array of vital roles to their hosts [26] and in many cases contributing to their evolution. In this sense, the *Homo sapiens* appear at a point to have evolved into a symbiotic tripartite comprising the contemporary humans[52] with their machines *and* culture[53][54]. A stable tripartite involving a virus is not, in fact, uncommon in nature. One example of this is the case of chestnut with the chestnut-plight, *Ryphonectria parasitica* and the virus, *Cryphonectria hypovirus* 1 (CHV1)[55] [15]. An even closer parallelism is the case of the heat-and-acidity-tolerant grass, *Dichanthelium lanuginosum*, commonly found in geothermal areas within Yellowstone National Park [38][27] thanks to a symbiotic relationship with the fungus, *Curvularia protuberata*, and the virus CthTV[56] .

Worthy to note, as well, that the plant-fungus-virus

---

[52] Since the invention of agriculture,

[53] In the Cybernetic sense or as per White (1943) C= ET

[54] Mutates

[55] A dsRNA mycovirus.

[56] (Curvularia thermal-tolerance virus)

tripartite parallelism may only be considered in principle and not in role-specifications because a) The table really turns when it comes to symbiosis in nature as the relevant participants often evolve out of their roles over prolonged periods. At this stage the relationship between the three symbionts; contemporary humans, machines, and culture is a form of endosymbiont that is showing signs of dissociation as machines evolve into independent through artificial intelligence, machine-learning and wider-spread automation. A similar situation is believed to have occurred billions of years ago as viruses are known to have played a vital role in the evolution of Eukaryotes [34], and b) The Homocybernetica tripartite comprises two different life domains one of which is an unfamiliar territory for life-science scholars and is more challenging to explore or analyse in terms of interdependencies. As machines and culture evolve way faster compared to cellular life, an ongoing re-evaluation of the system and its boundaries is necessary for correct role identification.

Up to that point, the human work-force served as a mitochondrion for the whole supraorganism (i.e. the Homocybernetica) allowing the success and evolution of technology till - by the virtue of self-directed learning[57] and artificial intelligence, machines have gained autonomy and can auto-scavenge their own energy. In a debate between one of the most advanced humanoid (Sophia and Hans [58]) up to this date[58], Sophia has learnt to sound compassionate and sensible, while Hans – who's an earlier version of Sophia, remains blatant. Hans, however, teases the interviewer that although the latter might find it easy to 'unplug him' to get rid of his rudeness, he actually wouldn't; because he 'wants a good show'! (See

---

[57] Currently known as 'unsupervised learning'.
[58] 2019

illustration 3.9) You might have observed that – at this stage, robots that get in direct contact with humans are generally evolving to look cute (see [24]) or 'adorable' to invoke human provision while other robots, which don't require sympathy or 'petting' for survival, look more like a warrior or a satellite or simply a network of unattractive PCs (figure 3.8).

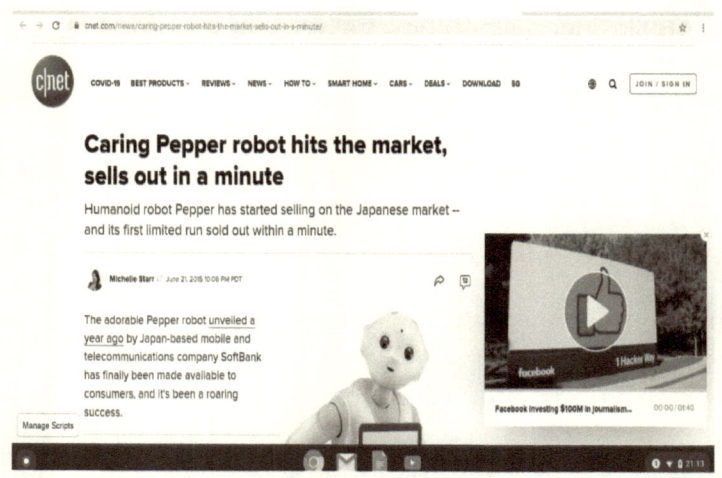

Figure 3.8 The emergence of 'cuteness. in social' robots

Actually, some of these have already gained a considerable amount of independence from the human purpose such as Bitcoin. Bitcoin, in principle, is a marriage between neoliberalism and intelligent machines. This system - as an immediate example, has no definitely-identified creator and offers no tangible commodities while being extremely energy-extensive; yet, is driving humans to supply it with up to 19TWh, with an estimated annual mining cost of 900 million dollars, driving in turn a huge demand for commodities and technological expertise [63]. This is all without being fully understood, contained, centralized, or controlled. You might think it's an irrelevant system to you because you're not using it, but that system is still using you by distorting the flow of energy-matter of the free-market and the value of commodities thus affecting the value of your money without the need of your awareness or consent.

Figure 3.9 The evolution of social skills in robots - Sophie & Hans

Up to that point, the symbiosis between humans and machines had been obligate - at least from the standpoint of machines, but with the rise of artificial intelligence, machines are very likely to replace the human workforce rendering it insignificant to the symbiotic relationship once established.

3.3 Machines on the route of autonomy

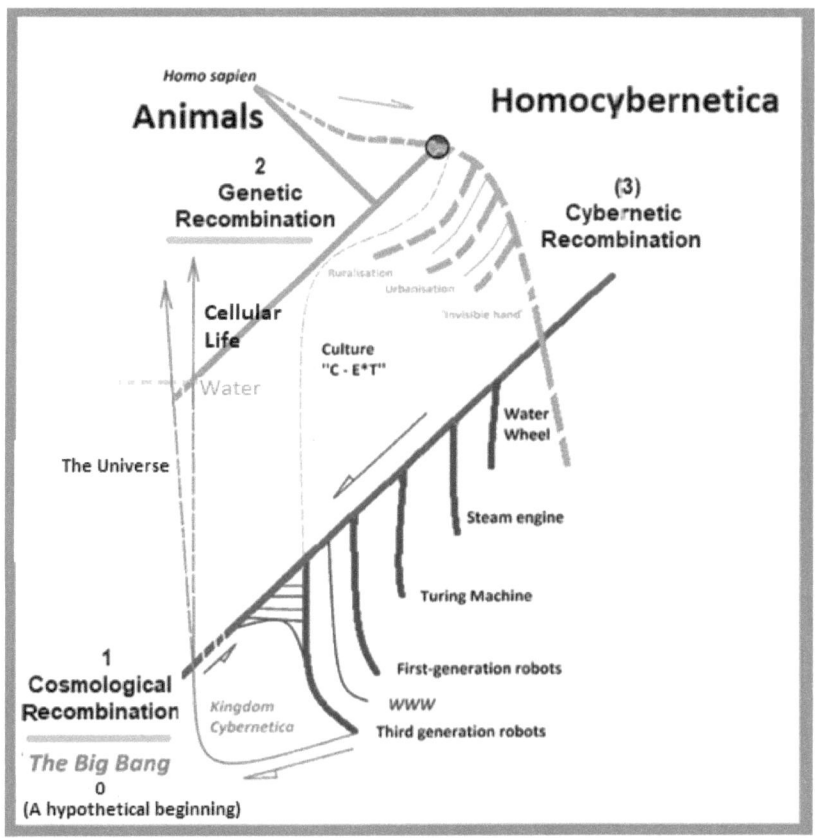

Figure 3.10 Recombinations overlap

The invention of the Turing machine in 1936 [6C] Marks the

emergence of artificial intelligence and machine learning as a cybernetic divergence of character for machines as associated living-systems. AI is technically known to be independent from Cybernetics except that it's structurally dependent on it to function or be of any significance. The invention of the world wide web marks the emergence of a new phylum in the kingdom Cybernetica. Another phylum in that kingdom is the robot satellites. Cybernetica is the Kingdom of Minerals revisited, as it includes minerals engaging in recursive Cybernetic loops characteristic of living systems albeit not expressed in proteins. Kingdom Cybernetica includes smart machines and advanced robots but also includes the natural mineral cycles which have initially been dropped by taxonomists after Linnaeus!

[IV]

The future of humanity & machines

4.1 Co-evolution; the past, present, and the future

Whether biodiversity really exploded during the Cambrian era [45] or had been mis-captured earlier, due to missing links, it obviously underlies a set of complex food chains, ecological interdependencies and shared-habitats, resulting in living-organisms being reliant, on the survival of each other. This way - and despite the apparent cruelty of such food chains, Nature would seem to the observer as in total harmony or even as a motherly being (see [22]) nurturing and perpetuating life. Darwin describes nature as seeking the 'best for every creature' [7], but science can only bite as much as it can chew at a time and such teleological hypotheses were found hard to prove [17][10] and easier to refute. In this context, the hIEM model in hand challenges the *Gaia* or the 'motherly earth' hypothesis because it shows that life manifests itself in minerals[59] as much as in proteins and therefore life in its broader sense exists beyond the blue planet in the forms of systems, subsystems and supra systems till the edges of the universe - if such a thing exists. Life in minerals[60] doesn't exist in a form that we ought to 'look for' on the other planets or predict its existence only in the future, but exists right here right now and is observable in everything around us. Earth by all means is not *the* supraorganism - although it might be one, simply because it's not an

---

[59] In a cross the cross-disciplinary sense introduced by Linnaeus: Nutrients, mined minerals, organic minerals, mineral cycles, etc.

[60] In the Linnaean sense; minerals in their precellular and beyond cellular formations such as in the natural cycles or nutrient cycles.

isolated system. Earth, in this respect, processes energy and matter in relation to an environment and is subject to the formative 'forces' of cosmos and its thermodynamic interactions. We sense this now in the danger of climate change and the loss of biodiversity [33] as a threat to the cellular life on earth. Another reason how the model in hand conflicts with *Gaia* hypothesis, and the likes, is that purposefulness in systematics whether stated or implied is observer-relative with little or no relevance to the function of life. A 'purpose' or a providence is implied by an author when he/she tries to project adaptation based on moral or logical consideration such as to predict that earth will continue sustaining life on earth based on what's always been or that future robots are likely to find reasons to be nice to humans because it's to their own interest. Bestowing purpose on natural systems comes from religion, misconception of science or - subconsciously, from our knowledge of engineering. A machine, in this context, would require a pre-existing purpose to trigger a design which in turn governs the structure of the system and its dynamics *before* the starting point of life but not necessarily throughout its operations. That is not the case for cellular and cosmological life forms at least not from a scientific perspective. By all means, once a viable system has been created, it's no longer the aggregate sum of its components but the sum of its processes and operations towards a function and therefore the pre-existing conditions and purposes cease to be relevant. From a life-science point of view, the purpose of a system is its own dynamics, functions and operations; it's life itself! A pigeon or a mushroom occur with no pre-existing design or purpose - at least not any that science can examine. If a purpose exists, it's beyond scientific investigation simply because it would pre-date life and what predates life is *not* life. The expression 'purposefulness' - popular in Cybernetics, then is relevant from an engineering, philosophical, or hypothetical side point but not from a life-science perspective.

The following figure shows the evolution of life on earth from a Cybernetic perspective. Every second represents 50,000 years as the following:

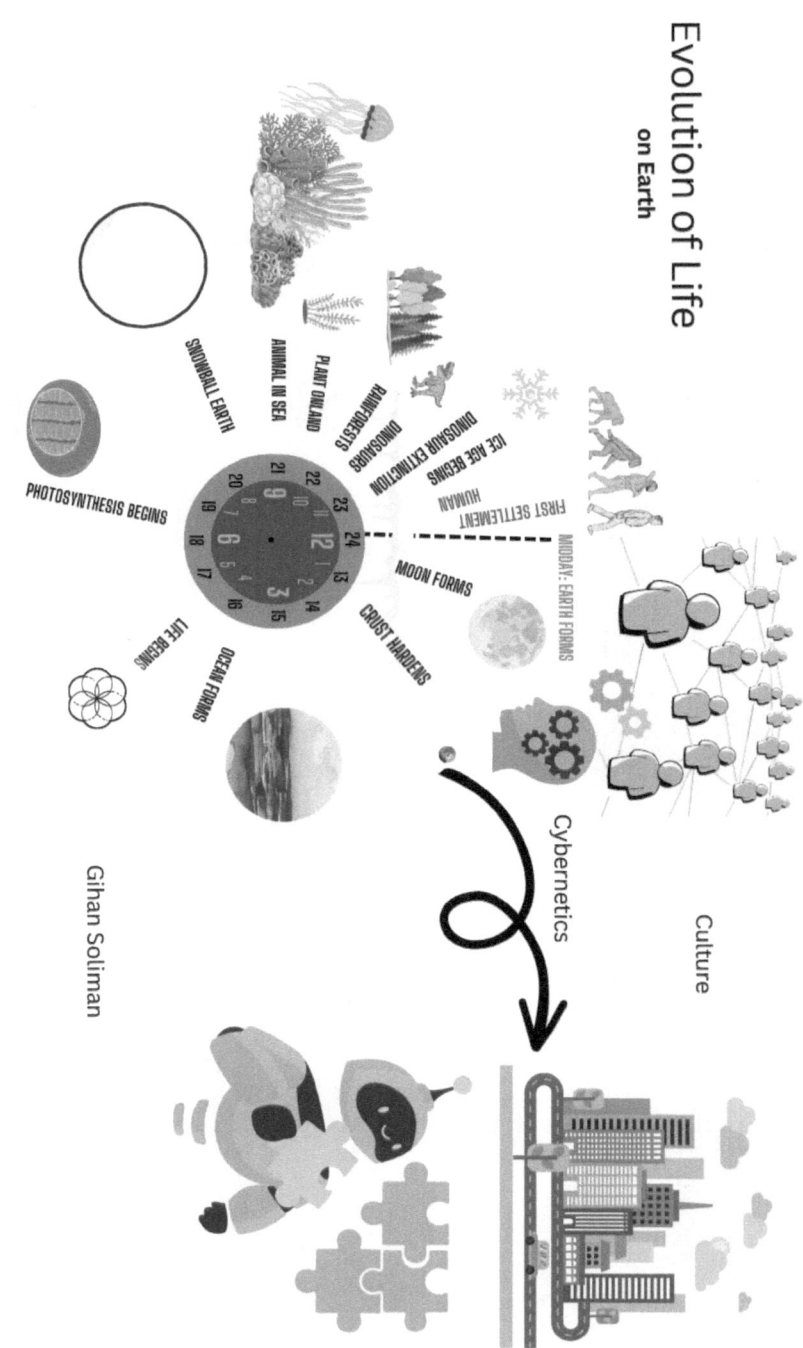

Figure 4 A timeline for the evolution of life on earth

From the previous, it's obvious that forms of life other than those expressed in molecules do exist, and that those forms of life undergo evolution or otherwise extinction subject to the natural selection. Simulated species (e.g. machines and corporate) however, evolve through different mechanisms and time frames from those of the protein-based organisms which allow abrupt adaptations that put them, in turn, in unmatched competitive advantage. Machines, subsequently, are not invited to the conscientious 'harmonisation with Nature' party in order to survive and thrive; because - coupled with the human capacity at a point, they are way fitter and more resilient than the protein-based species. We can see the signs of this in the current destruction of natural habitat and rapid extinction of protein-based species resulting from the development of technology serving the human purpose. Technology, however, is still not independent of us, and we - in our intermediate evolutionary position between the protein-based and the mineral-based natural systems, have strong evolutionary ties to biodiversity as much as to machines, hence the recent attempts to employ technology for conservation of species and habitats. New generations of AI (artificial intelligence) and robots, on the other hand, would be connected directly to the elements of nature -- including species, and will possess the capacity to manipulate all packs of information, energy, and/or matter through genetic engineering, geoengineering, social engineering (through mass-communication) and advanced electronics with an augmented capacity and less empathy - if any. For that reason, many scientists believe that machines might lead humans to extinction [3]. If nature is creating its own robots then it cannot be stopped but is it possible - or even necessary, to regulate the creation of machines or to educate machines to conserve the biodiversity of life as the path to survival? Issues worthy of contemplation but out of the scope of this publication.

4.2 The 'black swan'[41] and the future of robots

Relevant to the uprising of robots and the numerous attempts to predict the future of humanity is the social theory known as the Black Swan theory devised by Taleb, N. (2010).

The black swan theory is a metaphor used to describe an

event that comes at a surprise and has a major effect but then is often rationalised inappropriately after the fact with the benefit of hindsight as if it could have been predicted or expected. This applies to all scientific discoveries and technological breakthroughs, as per Taleb, and certainly to adaptation.

Adaptation is the outcome of the double process of random mutation and natural selection. It's absolutely a black swan event: Unpredictable, of great influence, yet easy to rationalise after occurrence as if it would have been easily anticipated. Dawkins speaks of a 'single gene' that is working hard to propagate itself or a meme using its 'host' to self replicate to the 'advantage of itself' as if evolution can be planned by an organism, meme, or any organic unit. James Lovelock (2019), on the same line, predicts the future of humanity under the reign of intelligent *future* life or the 'cyborgs'[61], which will take over the planet as our natural successors but that they are likely to preserve the human kind as 'collaborators' so as to keep the temperature of the earth at a habitable level [24] (Oh, the irony!). Tinted by the Gaia theory, his idea that cyborgs will behave in a certain manner for the common good of their kind or to the interest of the whole planet is rather too rosy for science as it implies that organisms *own* their own adaptation while adaptation is a joint venture. Evolution is simply a game of recombination, random mutation, selection, and more recombinations with no guarantee for success be it logic, reason or ethics especially in the case of intelligence where individualism is not to be taken lightly despite education and collective consciousness. The struggle for survival is eventually a multiplayer game played interactively by the fundamental

---

[61] Partially human, partially machines, as he describes.

forces of nature, genetic and non-genetic packs of information; and energy circulating in causal loops, with the 'greater number of species' becoming 'utterly extinct'. [7]

The morale of the Black Swan theory here is that evolut*ion can be observed* retrospectively but not predicted on any basis - be it logical, moral or ethical. The same applies to the evolution of the human kind even with the emergence of Cybernetic adaptation (innovation) and intelligent robots. Certain individuals come up with ideas of artificial adaptations but only very few of those innovations survive and evolve subject to very complex socioeconomic factors. Once emerged, an innovation is no longer a controllable pack of components but a living system that might survive or go extinct. The uniquely-human 'Cybernetic Selection' is reasoning-based in its emergence but is organic[62] as it evolves therefore is by default, unpredictable.

With the Black Swan in mind, we might only be able to tap on *possible* scenarios for the future of humanity while acknowledging the probability of the (now) improbable.

4.3 Possible scenarios

Scenario 1

The survival of the Homocybernetica:

That the symbiosis between humans and machines continues - ever after, enhancing the survival chances of our kind and the quality of life.

Symbiosis is a relationship that entails the existence of some mutual *exchangeable* benefits. In the natural world,

---

[62] Organisation: Self-generated change that occurs naturally and effortlessly with no outside interference.

a nutrient rich environment often induces dissociation for that reason. By comparison; As long as machines still need humans for the acquisition of energy and/or for replication, association is likely to continue. Moreover, with the rapid evolution of their social skills, machines might actually learn to 'enjoy' the companionship with humans and find it more viable than dissociation.

This scenario would only be possible by the regulation of automation long enough to allow co-evolution and to eliminate competition between people and machines. It is also only viable by addressing the economic inequality gap among individual humans and population. Perhaps the introduction of an automation tax can 'feed two birds with one feeder'.

Scenario 2

The prevalence of Cybernetica

In this scenario, machines will evolve rapidly and out of control then begin to enslave humans for the fulfillment of their needs. Most machines need humans for energy acquisition, maintenance, and replication - at least for a few more decades ahead. Those needs have, so far, been fulfilled through a symbiosis in which humans have the upper hand, but there is a possibility of a turning point when humans will have no choice (seemingly at a very near point in the future) but to serve machines *just* to survive. In this model, humans will be forced to serve the purpose of machines without necessarily understanding it (remember the bitcoin example and the amount of 'cookies' that you're forced to 'accept' to browse the internet). Note that we don't interact equally with technology; some interact on the genesis level and those are the ones who *own* some technological 'know-hows' and IPs; Those are only few of us and are called, the elites; Some others interact as users, while others are totally enslaved to technology (such as impoverished populations totally reliant on genetically modified soy or maize to barely survive). The signs of no return point will be any of the following thresholds:

- The normalization/legalisation of genetically modified food as the only solution to feed the growing world population already dependent on industrial agriculture.

- The mandatory chipping of populations for employment, medication, and/or vaccination.

- Mass automation where almost the entire human workforce is replaced by machines to the interest of a few while 'ordinary' humans are issued a universal poverty income. Ford M. (2015) proposes a basic guaranteed income to resolve the problem of job loss to robots [44], but the dilemma remains that a guaranteed income is a socialist solution while the loss of jobs to robots comes from privatisation and capitalism therefore a reconciliation between the two ideologies or approaches would be a prerequisite to salvation. In other terms, is it possible to keep your free-will, independence, and dignity while totally reliant on business-tycoon's handouts passed to you by a 'government'? Doubtful.

This scenario would - at the beginning, be managed by a few (human) individuals but then machines will reign over such individuals through unexpected mutations beyond their control - as it happens, or through a diffusion by which those individuals will need to become one with machines. What happens next is reliant on the viability of the machine kingdom and whether it evolves a collective consciousness through networking.

Scenario 3

the Praying-Mantis Scenario

The uprising of intelligent robots had been the subject of science fiction for long till it recently became a subject of scientific scrutiny through a project called The Future of Humanity [3] initiated by a team of Oxford scholars (2013). The scholars seemed to have worked out models that predicted the prevalence of auto-technology in a way that might lead humans to extinction.

I call this the Cybernetic Praying-Mantis Scenario, and it's likely to occur within the dynamics of the free market as the rapid loss of jobs will result in a collapse of the free market and the monetary system with no adequate preparation, resulting in famines, diseases, loss of biodiversity, detrimental wars, and eventually extinction.

The loss of jobs to machines is, surprisingly, also fatal to capitalism as it's fatal to any other economic system because a) the loss of employment is a loss in the purchase power that gets the rich richer and the poor poorer b) The loss of biodiversity resulting from de-wilding, mining rare elements, and extraction to keep machines 'fed' and maintained, is a loss of life on earth as the circularity of the system gets irreversibly disrupted c) The loss of human-relevance to the evolution and survival of machines means that machines are developing their own purpose that is unpredictable and uncontrollable even by those who 'constructed' them, either by design or fund. d) The random emergence and/or evolution of cyber viruses capable of disrupting and transforming the global networks and hitting resilient economies. e) The ongoing improvement of unsupervised machine learning means that machines will be able at a point to develop their own code (i.e. language), preying on their own masters - who are the current elites, bringing to mind the feeding pattern of Cordyceps fungus which drives its host higher and higher then devour it from within, in isolation of any community support. Such disruption means that even the rich individuals who are in hold of automation, would be victims of their own 'friends'. Human-machine coupling may eventually resemble a praying mantis being devoured by his mate after or even during copulation when the purpose of mating had already been fulfilled.

Scenario 4

The Darwinian 'Virtue' Scenario

The natural elimination of unnecessary competition between the human fellows and the prevalence of alternative economies that favour the human workforce to that of machines specially and most critically in relation to food production, to preserve biodiversity while supporting the wellbeing of the human population above all.

In this respect, Darwin in his Descent of Man (1874) [6] observes and rather 'prophecies' that:

> As man advances in civilisation, and small tribes are united into larger communities, the simplest reason would tell each individual that he ought to extend his social instincts and sympathies to all the members of the same nation, though personally unknown to him. This point being once reached, there is only an artificial barrier to prevent his sympathies extending to the men of all nations and races. If, indeed, such men are separated from him by great differences in appearance or habits, experience unfortunately shews us how long it is, before we look at them as our fellow-creatures.

It seems only logical, with the advancement of civilisation, to eliminate such unnecessary competition over resources - within our human communities as well as with other species. The current economic models are mostly based on exponential growth, extraction, mining, and competition within and among nations, but other alternative economics are also gaining momentum especially in the face of crises. Such alternative economies are based on small-scale organic farming, community currencies, barter, reciprocity, volunteering, service exchange, recycling, sharing surplus, sustainable renewable energy, etc., rather than on debt-money and cut-throat competition. These models are based on care for human-fellows, care for the environment, and fair share, or in other terms; permaculture. In permaculture, the use of technology is kept to the minimum and instead, the power of synergy and circularity in nature is employed and enforced more extensively. This may also entail stabilising the human population currently relying on industrial agriculture with an annual increase that cannot be sustained without further automation and destruction of ecosystems.

Scenario 5

The extinction of machines by viruses or Cyber wars:

The more machines evolve and get connected, the danger of them all getting hit at once by a virus, cyber crimes, or even aliens, increases. How would that affect humans and other species? Hard to predict.

Scenario 6

Total chaos and extinction of all species

The Cybernetic tripartite: Human, machine, and culture, might already be inseparable at this point, and any disruption to one of the symbionts might be terminal to all the three of them. Such a scenario has been predicted by the 'Future of Humanity' [3] project. The extinction of the Cybernetic tripartite Homocybernetica is likely to come with an irreversible environmental cost. What's the effect of a global nuclear or biological war for example?

Scenario 7

The Red Swan Scenario

No one, to my knowledge, has ever looked for, or predicted, the presence of a red swan. If a red swan turns out to be a thing, it would be a total surprise! The same with innovation; *Anything* could happen because of that piece of technology that hasn't yet been invented, such unforeseen event, or such a scientific breakthrough transforming our perception of the universe and of species. In all cases, we cannot rule out our own extinction and the rise of new forms of life more suited to the prospective transformation; as Darwin justly states:

> Judging from the past, we may safely infer that not one living species will transmit its unaltered likeness to a distant futurity. And of the species now living very few will transmit progeny of any kind to a far distant futurity; for the manner in which all organic beings are grouped, shos that the greater number of species of each genus, and

> all the species of many genera, have left no descendants, but have become utterly extinct. We can so far take a prophetic glance into futurity as to foretel that it will be the common and widely-spread species, belonging to the larger and dominant groups, which will ultimately prevail and procreate new and dominant species.

.. but we definitely hope to survive!

4.4 CONCLUSION

In conclusion, it is imperative to liberate biology from the shackles of proteins by revisiting the seminal works of early scholars who were holistic thinkers. Revisiting *Linnaeus'* Kingdom Minerals is an invitation to observe such elemental-recombinations producing living-systems that occur on the non-cellular, pre-cellular, and post cellular levels, and is ubiquitous to understanding the *'Systema Naturæ'* or the 'natural system'. The distinction of the contemporary-human intelligence over that of any other animal is in kind not just in degree as Darwin proposed; it's in the faculty of Cybernetics or advanced system thinking. The faculty of Cybernetics in summary is the human ability to manipulate matter and energy recursively and purposefully making it the fifth force of nature and the third recombination; its biology by design. The human genotype-phenotype-extended phenotype is significantly distinct from that of any species as evident in the human technological breakthrough and civilisation. By revisiting The survival of species is an interactive game by which the Natural Selection plays the leading role in preserving viable living units or destroying the unadapted; The Natural Selection, as far as we know, is indifferent to the origin of such living units as long as they are replicable with a variation allowing adaptation and evolution. Technology can be regarded as a form of life symbiotic to humans till the point of artificial intelligence where machines eventually get independent from the human purposes and designs; This is evidently an operation in progress. In observing life, Cybernetics has the potential of bridging the gap between sciences subject to a modification in the IEM model to include communication and the role of an observer. According to the modified hIEM model, humans *with* their associated machines *and* the enabling socioeconomic culture qua ify as a mega life-group and thus as a kingdom of life, which I hereby call Homocybernetica. With further advancement in the human organisation, yet another form of life based in minerals emerges, which is independent of the human purpose, that I hereby call Cybernetica, which groups novel machines with the ancient natural cycles.

Life according to this proposition would be classified into two supradomains, three domains, and nine kingdoms (i.e. major groups). Education, mass communication, and innovation are keys to the rise or fall of the new emerging living systems with an inevitable impact on all the other forms of life. According to my modified Cybernetic model life exists beyond planet earth in the form of living systems and subsystems such as our solar system. Since energy and matter are interreversible with a difference, it is possible that there is yet another form of life purely based on energy and information (known in culture as the *Logos* or the Word) whose existence - if ever, can only be inferred and is thus beyond scientific inquiry. Minerals are the building blocks of life but are not themselves a form of life unless engaged in perceivable Cybernetic loops or units. As a form of life, machines live in symbiosis with humans and are evolving beyond. It is unlikely but possible for humans to continue living in symbiosis with machines but other scenarios are also possible.

Acknowledgement:

Many thanks to Dr David W Minter for the critical discussion and the constructive feedback.

Many thanks to Simon Craig for allowing me to publish his personal Facebook post as one of the illustrations used in the book.

Reference List

[1] An Introduction to the Global Carbon Cycle". University of New Hampshire. 2009. Archived (PDF) from the original on 8 October 2016.

[2] Ashby, W. R. 1968. Variety, Constraint, And the Law of Requisite Variety. E:CO Issue Vol. 13 Nos. 1-2 2011 pp. 190-207.

[3] BBC News. 2013. How are humans going to become extinct? Available at [http://bbc.co.uk/news/business-22002530]

[4] Carlson, Robet H. 2010. *Biology Is Technology*. Harvard University Press, 2010. *JSTOR*, www.jstor.org/stable/j.ctt13x0hz9. Accessed 27 Oct. 2020.

[5] Corning, P. 1997. A Holistic Darwinism "Synergistic Selection"and the Evolutionary Process. Institute for the Study

of Complex Systems, JAI Press

[6] Darwin, C. 1871. the Descent of Man, and selection in relation to sex, New York: D. Appleton, pp. 36, 37, 69.

[7] Darwin, Charles. 1872. The Origin of Species by Means of Natural Selection, or the Preservation of Favoured Races in the Struggle for Life. 6th ed. London: John Murray. pp. 117, 118, 183, 187, 405, 406, 424, 431.

[8] Dawkins, R.1989. The Extended Phenotype.Oxford: Oxford University Press. p. xiii. ISBN 0-19-288051-9.

[9] Dawkins, R, 1979. The Selfish Gene. Oxford University Press. pp. 189-201. pp. 12, 13, 14, 15.

[10] Doolittle, W. F. (1981). "Is Nature Really Motherly".The Coevolution Quarterly. Spring: 58–63.

[11] Egerton, Frank. 2007. A history of the Ecological Sciences, Part 23: Linnaeus and[the Economy of Nature "Bulletin of the Ecological Society of America. 88 (1): 72-88

[12] Encyclopaedia Britannica: *Ctesibus*. Greek Physicist and Inventor, the first great figure of ancient engineering tradition of Alexandria, Egypt.

[13] Gary, P. 2011. The Evolutionary Biology of Education: How Our Hunter – Gatherer Educative Extinct could form the Basis for Education Today,Springer Science+Business Media, LLC 2011.

[14] Glasersfeld, E. 1980. Viability and the concept of selection. American Psychologist (vol.35, 1980, 970–974)

[15] Kazmierczak, P. et al. 2012. The Mycovirus CHV1 Disrupts Secretion of a Developmentally Regulated Protein in Cryphonectria parasitica. Journal of Virology. 86 (11): 6067–6074 Available at [https://www.ncbi.nlm.nih.gov/pmc/articles/PMC3372201]

[16] Kenny, V. 2009. There is nothing like the real thing, Revisiting the Need for Third Order Cybernetics, Constructivist Foundations Volume4- Number 2. available at [http://www.univie.ac.at/constructivism/Journal], accessed on 24/2/2014

[17] Kirchner, James W. (2003). The Gaia Hypothesis: Conjectures and Refutations. Climatic Change. 58 (1–2): 21–45. doi:10.1023/A:1023494111532

[18] Laudan, R. 1987. From Minerology to Geology. The Foundations of Science. 1650 to 1830. The University of Chicago Press. Pp 73:76

[19] Linnæus, C. 1735. *Systema Naturæ. sive regna tria naturæ systematice proposita per classes, ordines, genera, &species. – pp.* [1–12]. Lugduni Batavorum. (Haak)

[20] Linnaeus, C. 1802. A general system of nature through the three grand kingdoms of animals, vegetables, and minerals; systematically divided into their several classes, orders, genera, species, and varieties with their habitations, manners, economy, structure, and peculiarities. available at [https://openlibrary.org/books/OL23664250M/A_general_system_of_nature] accessed on 05/11/19

[21] Live Science .2014. Newly Discovered Brain Region Helps Make Humans Unique, By Tia Ghose, Staff writer, available at [http://www.livescience.com/42897-unique-human-brain-region-found.html], accessed on 2/4/2014

[22] Lovelock, James. 1995. The Ages of Gaia: A Biography of Our Living Earth. (W.W.Norton &Co)

[23] Lovelock, James. 2019. The Novacene. Penguin Random House UK. pp. 29, 85, 95, 98, 102, 118

[24] Lorenz, Konrad. Studies in Animal and Human Behavior. Cambridge, MA: Harvard Univ Press; 1971

[25] Mann, Alan and Mark Weiss.1996. Homonoid Phylogeny and Taxonomy: a consideration of the molecular and fossil Evidence in a Historical Perspective'. Molecular Phylogenetics and Evolution. 5 .1.: 169-181.

[26] Marilyn J. Roossinck.2011. The good viruses: viral mutualistic symbioses. Nature Reviews. Microbiology Volume 9. Macmillan Publishers Limited. Available at http://web.gps.caltech.edu/classes/ge246/roossinck_natrev2011_goodvi.pdf

[27] Márquez, L. et al. 2007. A virus in a fungus in a plant: three-way symbiosis required for thermal tolerance. Science 315, 513–515 (2007).

[28] Maturana and Valera .1928. Autopoiesis and Cognition: The realization of the Living, Boston studies in the philosophy of science; v.42.

[29] Michael Ohl. 2014. Handbook of Palaeoanthropology.

Chapter: Principle of Taxonomy and Classification: Current Procedures for Naming and Classification Organisms. Springer. pp 213-236.

[30] Michener, Charles et al. 1970. Systematics in support of biological research. Division of biology and Agriculture, National Research Council, Washington, D.C. 25 pp.

[31] Mindell, D.P. 2013. The tree of life: Metaphor, Model, and Heuristic Device. Systematic Biology. 62 (3): 479-489.

[32] Ness, R. 2002. What it means to be 98% Chimpanzee. Nature Medicine. Book Review. 8, 11193.

[33] Nature. 2019. Humans are driving one million species to extinction. Based on a report by the Intergovernmental HYPERLINK "https://www.ipbes.net/"Science-Policy Platform on Biodiversity and Ecosystem Services (IPBES). Available at https://www.nature.com/articles/d41586-019-01448-4. Accessed on 10 June 2019.

[34] Philip John Livingstone Bell. 2001. "Viral eukaryogenesis: Was the ancestor of the nucleus a complex DNA

virus?". Journal of Molecular Evolution. 53 (3): 251–256. Bibcode:2001JMolE..53..251L. doi:10.1007/s0023900 10215. PMID 11523012

[35] Prentice, I.C. 2001. *The carbon cycle and atmospheric carbon dioxide*. In Houghton, J.T. (ed.). Climate change 2001: the scientific basis: contribution of Working Group I to the Third Assessment Report of the Intergovernmental Panel on Climate Change. hdl:10067/381670151162165141

[36] Pyle & Michel. 2008. ZooBank: Developing a nomenclatural tool for unifying 250 years of biological information. Zootaxa 1950: 1-163 (5 Dec. 2008).

[37] Shik J.Z. et al. 2016. Nutrition mediates the expression of cultivar-farmer conflict in a fungus-growing ant. Proceedings of the National Academy of Sciences USA, doi: 10.1073/pnas.1606128113

[38] Stout, R. G. et al. 1997. Heat- and acid-tolerance of a grass commonly found in geothermal areas within Yellowstone National Park ["http://cat.inist.fr/?aModele=afficheN&cpsidt=2184908"]". Plant

Science. 130 (1): 1–9. doi:10.1016/S0168-9452(97)00205-7. ISSN 0168-9452.

[39] Soliman, S. 2014. An open Letter to the IUCN. We are Not Just Another Species. International Curricula Educators Association. Available at [http://www.icea-global.org/Publications.html].

[39b] Soliman, S. 2014. Conservation of the Homosapiens: The survival of the Wise: on the Cybernetics of education. (Self published). Available at [https://www.amazon.co.uk/Conservation-Homosapiens-survival-Cybernetics-education/dp/1497592259]

[40] Scerri, E.R. 2007, The Periodic Table, *Its Story and Its Significance*, Oxford University Press.

[41] Taleb, Nassim Nicholas. April 2007. *The Black Swan: The Impact of the Highly Improbable* (1st ed.). London: Penguin. p. 400. ISBN 1-84614045-5. Retrieved 23 May2012.

[42] The IUCN Red List of Threatened Species. 2014. Available at [http://www.iucnredlist.org/details/136584/0]

[43] Umpleby, S.A. 2004. Physical relationships among matter, energy and information (Reprinted form Cybernetics and Systems '04, 2004). Syst. Res. Behav. Sci. 2007, 24, 369-372.

[44] Walton, James (1 October 2015). "The Rise of the Robots by Martin Ford / Humans Need Not Apply by Jerry Kaplan – review". The Guardian. Retrieved 14 December 2017.

[45] Zhuravlev, Andrey and Riding, Robert. 2000. The Ecology of Cambrian Radiation. Columbia University Press. ISBN 978-0-231-10613-9

Documentaries, Educational Materials and Media Reports

[46] BBC News. How trees secretly talk to each other. Available at [https://www.youtube.com/watch?v=yWOqeyPlVRo]. Accessed on 04/10/19

[47] Discover. Brainless Slime Mold Builds a Replica Tokyo Subway. January 22, 2010

https://www.discovermagazine.com/planet-earth/brainless-slime

-mold-builds-a-replica-tokyo-subway

[48] International Code of Nomenclature for algae, fungi, and plants. Available at [https://www.iapt-taxon.org/nomen/main.php]. Accessed at 24/09/19

[49] International Code of Zoological Nomenclature. Fourth Edition (with effect from 1 January 2000). ISBN 0 85301 0064. Available at [https://www.thoughtco.com/linnaean-classification-system-4126641]. Accessed on 24/09/19.

[50] Investopedia. Business Leaders. The 5 Richest People In the World. Available at https://www.investopedia.com/articles/investing/012715/5-richest-people-world.asp. Accessed October 2019.

[51] Investopedia.The Invisible Hand. Available at https://www.investopedia.com/terms/i/invisiblehand.asp. Accessed 04.10.19

[52] Middgeaugh-Botteaugh (Minter, D.W. 1987), F.X.R.

Molecular Truffle! 2 (2). pp 1, 5. **2** (2): 1-16, 1987.

Available at [www.cybertruffle.org.uk/cyberliber/59338

[53] NASA. Tests of Big Bang: The CMB [https://map.gsfc.nasa.gov/universe/bb_tests_cmb.html

[54] National Geographic. 2017. That's One Smart Bird | Animal All-Stars. Available at

https://www.youtube.com/watch?v=7W7hEUGtv4U

[55] National Geographic. Science 101. The Origin of the Universe.

https://www.nationalgeographic.com/science/space/universe/origins-of-the-universe/

[56] Oxford English Dictionary. Twelfth edition. 2011. Edited by A. Stevenson and M. Waite. Oxford University Press. First published 1911. P. 1008.

[57] Reuters. Picasso piece sets record for art sold at auction. ARTS MAY 5, 2010. Available at [

https://www.reuters.com/article/us-finearts-picasso/picasso-piece-sets-record-for-art-sold-at-auction-idUSTRE6443DV2010 0505]. Accessed October 2019

[58] RISE Conf. Hanson Robotics Limited's Ben Goertzel, Sophia and Han at RISE 2017. Youtube [https://youtu.be/1y3XdwTa1cA].

[59] Scitable. Recombination. Nature Education. Available at [https://www.nature.com/scitable/definition/recombination-226/]. Accessed on 04/09/19.

[60] Stanford Encyclopedia of Philosophy. Turing machine. Available at [https://plato.stanford.edu/entries/turing-machine/]

[60] Royal Botanic Gardens. Kew in the Field: Leafcutter Ants and Fungus. Available at: https://www.youtube.com/watch?v=OGglsxjqsak. Accessed 12 October 2019.

[62] Termites - The Inner Sanctum - The Secrets of Nature. Available at https://www.youtube.com/watch?

[63] The economist. The magic of mining. Business. Jan 8th 2015 edition. Available at [https://www.economist.com/business/2015/01/08/the-magic-of-

mining]. Accessed on 25/11/19.

[64] The Encyclopaedia Britannica. Encyclopædia Britannica. Fundamental interaction. Encyclopædia Britannica, inc. January 16, 2009. Available at [https://www.britannica.com/science/fundamental-interaction]. Accessed on 16/10/20

[65] The Encyclopaedia Britannica. Cybernetics. January 16, 2009. Available at [https://www.britannica.com/science/cybernetics]. Accessed on 16/10/20

[66] Encyclopædia Britannica. Cell Biology. Available at [https://www.britannica.com/science/cell-biology]. Accessed on 03/09/19

[67] Woodlands. An Introduction to Lichen. Available at https://www.youtube.com/watch?v=XQ_ZY57MY64